Fred Piper and Sean Murphy

CRYPTOGRAPHY

A Very Short Introduction

OXFORD
UNIVERSITY PRESS

OXFORD

UNIVERSITY PRESS

Great Clarendon Street, Oxford OX2 6DP

Oxford University Press is a department of the University of Oxford.
It furthers the University's objective of excellence in research, scholarship,
and education by publishing worldwide in

Oxford New York

Auckland Bangkok Buenos Aires Cape Town Chennai
Dar es Salaam Delhi Hong Kong Istanbul Karachi Kolkata
Kuala Lumpur Madrid Melbourne Mexico City Mumbai Nairobi
São Paulo Shanghai Singapore Taipei Tokyo Toronto

with an associated company in Berlin

Oxford is a registered trade mark of Oxford University Press
in the UK and in certain other countries

Published in the United States
by Oxford University Press Inc., New York

© Fred Piper and Sean Murphy 2002

The moral rights of the authors have been asserted

Database right Oxford University Press (maker)

First published as a Very Short Introduction 2002

British Library Cataloguing in Publication Data

Data available

Library of Congress Cataloging in Publication Data

Data available

ISBN 0-19-280315-8

3 5 7 9 10 8 6 4 2

Typeset by RefineCatch Ltd, Bungay, Suffolk
Printed in Spain by Book Print S. L., Barcelona

LIBRARY

Tel: 01244 375444   Ext: 3301

This book is to be returned on or before the
last date stamped below. Overdue charges
will be incurred by the late return of books.

CHESTER COLLEGE

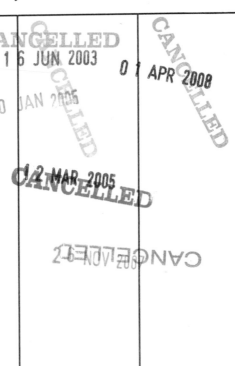

VERY SHORT INTRODUCTIONS are for anyone wanting a stimulating and accessible way in to a new subject. They are written by experts, and have been published in 25 languages worldwide.

The series began in 1995, and now represents a wide variety of topics in history, philosophy, religion, science, and the humanities. Over the next few years it will grow to a library of around 200 volumes – a Very Short Introduction to everything from ancient Egypt and Indian philosophy to conceptual art and cosmology.

Very Short Introductions available now:

Available soon:

For more information visit our web site

www.oup.co.uk/vsi

Acknowledgements

This book has evolved over a number of years with many people making valuable comments and suggestions. We thank them all. We are particularly grateful to Gerry Cole, Ross Patel, and Peter Wild for reading the final draft, and to Adrian Culley and Kalatzis Nikolas for saving us from embarrassment by correcting some of our exercises. Most of all we appreciate the contribution of Pauline Stoner who managed to convert apparent random scribble into a presentable format. Her patience was frequently tested and the book would not have been completed without her.

Contents

Chapter 1
Introduction

Most people seal the envelope before posting a letter. If asked why, then some immediate responses would probably include comments like 'I don't know really', 'habit', 'why not?' or 'because everyone else does'. More reasoned responses might include 'to stop the letter falling out' or 'to stop people reading it'. Even if the letters do not contain any sensitive or highly personal information, many of us like to think that the contents of our personal correspondence are private and that sealing the envelope protects them from everyone except the intended recipient. If we sent our letters in unsealed envelopes then anyone who gained possession of the envelope would be able to read its contents. Whether or not they would actually do so is a different issue. The point is that there is no denying that they would be able to if they wanted to. Furthermore, if they replaced the letter in the envelope then we would not know they had done so.

For many people the use of email is now an alternative to sending letters through the post. It is a fast means of communication but, of course, there are no envelopes to protect the messages. In fact it is often said that sending email messages is like posting a letter without an envelope. Clearly anyone wanting to send confidential, or maybe even just personal, messages via email needs to find some other means of protecting them. One common solution is to use cryptography and to encrypt the message.

If an encrypted message falls into the hands of someone other than its intended recipient then it should appear unintelligible. The use of encryption to protect emails is not yet particularly widespread, but it is spreading and this proliferation is likely to continue. Indeed in May 2001 a group of European MPs recommended that computer users across Europe should encrypt all their emails, to 'avoid being spied on by a UK–US eavesdropping network'.

Cryptography is a well-established science that has been a significant historical influence for more than 2,000 years. Traditionally its main users were governments and the military, although it is worth noting that *The Kama Sutra of Vatsyayana* contains a recommendation that women should study 'the art of understanding writing in cypher' (full details of all works cited are given in References and further reading.

The impact of cryptography on history is well documented. The tome on the subject is undoubtedly *The Codebreakers* by David Kahn. This book has over 1,000 pages and was first published in 1967. It has been described as 'the first comprehensive history of secret communication' and makes absorbing reading. More recently Simon Singh has written a shorter book called *The Code Book*. This is an easy-to-read account of some of the most significant historical events. It is not as comprehensive as Kahn's book, but is intended to stimulate the layman's interest in the subject. Both are excellent books that are highly recommended.

Credit for the popularization and increase in public awareness of the historical importance of cryptography is not restricted to literature. There are a number of museums and places of historic interest where old cipher machines are exhibited. High on the list of such venues is England's Bletchley Park, considered by many to be the home of modern cryptography and computing. It was here that Alan Turing and his team broke the Enigma cipher and their working environment has been preserved as a monument to their incredible achievements. Many recent films on the Second World

War have stressed the importance of code-breaking. Events that have received special attention are the impact of the breaking of the Enigma ciphers and the breaking of encrypted messages immediately prior to Pearl Harbor. There have also been a number of TV series devoted to the subject. All this means that millions of people worldwide have been exposed to the concept of encrypting messages to keep them secret and to the effect that breaking these ciphers can have. However, for many of them, the precise meaning of the terms used remains a mystery and their understanding is limited. The aim of this book is to rectify this situation by presenting a non-technical introduction to cryptology; the art and science of code-making and code-breaking. Readers will then be able to revisit these books, films, and TV series with extra knowledge which should make them more understandable and, as a result, more enjoyable.

Prior to the 1970s, cryptography was a black art, understood and practised by only a few government and military personnel. It is now a well-established academic discipline that is taught in many universities. It is also widely available for use by companies and individuals. There have been many forces that have influenced this transition. Two of the most obvious have been the move towards automated business and the establishment of the Internet as a communications channel. Companies now want to trade with each other and with their customers using the Internet. Governments want to communicate with their citizens via the Internet so that, for example, tax returns may be submitted electronically.

Whilst there is no doubt that e-commerce is becoming increasingly popular, fears about security are often quoted as being one of the main stumbling blocks for its complete acceptance. We have already focused on the problems associated with confidential information, but confidentiality is frequently not the main concern.

If two people are communicating over a public network and cannot see each other then it is not immediately obvious how either of them can establish the identity of the other. However, it is clear that anyone receiving a message over a network may need to be able to convince themselves that they know the identity of the sender, and that they are confident that the message they have received is identical to the one that the originator sent. Furthermore, there may be situations where the receiver needs to be confident that the sender cannot later deny sending the message and claim to have sent a different one. These are important issues that are not easy to solve.

In traditional non-automated business environments hand-written signatures are frequently relied on to provide assurances on all three of these concerns. One of the main challenges that security professionals have faced recently is to find 'electronic equivalents' to replace the social mechanisms, such as face-to-face recognition and the use of hand-written signatures, that are lost in the migration to digital transactions. Despite the fact that there is no obvious relation to the need to keep certain information secret, cryptography has become an important tool in meeting this challenge. In a 1976 paper that was appropriately entitled *New Directions in Cryptography*, Whitfield Diffie and Martin Hellman proposed a way in which cryptography might be used to produce the electronic equivalent to the hand-written signature. It is impossible to overstate the impact of that paper. Prior to their work, cryptography was being used to make users confident that their messages had not been altered during transmission. However, it relied on mutual trust between the communicating parties. This was not a problem for the financial institutions, which were probably the main users in the 1960s and 1970s, but environments where it could be employed were certainly limited.

Modern cryptography has evolved considerably over the past three decades. Not only has the technology changed, but there is a wider range of applications. Furthermore, everyone is likely to be either a direct user or be affected by its use. We all need to understand how it works and what it can achieve.

Using this book

This book provides an introductory overview of cryptography. It is non-technical and is written primarily for the layman. Mathematicians and computer scientists who wish to study the technical aspects of cryptography are already spoilt for choice. The basic theory of the design and analysis of encryption algorithms is well documented and there are numerous textbooks on the topic. (We regard the standard reference as being the *Handbook of Applied Cryptography* by Alfred Menezes, Paul van Oorschot, and Scott Vanstone.) This book is not meant to be another. It is not concerned with the technical issues associated with algorithm design, but concentrates on how algorithms are used and what they are used for. If it inspires readers with suitable mathematical backgrounds to read more specialized technical textbooks then it will have achieved one of its objectives. However, its primary objective is to try to remove the mystique that surrounds cryptography and to remove the fear with which many non-mathematicians regard it.

This book is based on a course in the M.Sc. in Information Security at Royal Holloway, University of London. The course was called 'Understanding Cryptography' but its title has been changed to 'An Introduction to Cryptography and Security Mechanisms'. The interests and backgrounds of the students on the course are varied but most of them have ambitions to become practising security professionals including, for instance, IT security managers or security consultants. Most of them do not wish to become professional cryptographers. In fact they enter the course regarding cryptography as a necessary evil that must be endured in order for them to obtain an Information Security qualification. While we, the authors, cannot regard the subject as 'evil', it is certainly true that cryptography should be studied within the context of providing secure systems, rather than as an independent subject in its own right. It is this attitude that justifies the assertion that it is generally more important for security practitioners to understand key management than to be able to analyse cryptographic systems mathematically.

For those who have no desire to be security professionals the aim of this book is to present cryptography as an interesting, important topic. It should enable the reader to understand the terminology in the numerous historical books and films on cryptography, and also to appreciate the impact cryptography has had on our history and is likely to have on our future. It should also facilitate understanding of the problems which the increased availability of cryptography causes for governments and law enforcement agencies.

There is little doubt that trying to break simple codes enhances one's understanding of cryptography. It can also be fun. So, although this is not a textbook, there are a number of 'exercises', in the sense that the reader is invited to break some algorithms. Failure to do so should not prevent the reader from completing the book. Nevertheless a serious attempt at solving them is probably worthwhile. The exercises are usually letter substitutions and solving the exercises requires no mathematics.

Despite the fact that there are essentially no mathematical prerequisites for understanding this book, there is no denying that modern cryptographic systems almost always involve mathematical processes. Furthermore most modern algorithms operate on binary digits (bits) rather than alphabetic characters. In recognition of this we include a short appendix to Chapter 3 with some of the relevant elementary mathematics. Once again readers are encouraged to try to understand them, but are assured that they are not crucial for the latter parts of the book.

Chapter 2
Understanding
Cryptography

Introduction

In this chapter we introduce the basic terminology and concepts of cryptography. Our aim is to be informal and to give as general an overview as possible.

The basic concepts

The idea of a cipher system is to disguise confidential information in such a way that its meaning is unintelligible to an unauthorized person. The two most common uses are, probably, to store data securely in a computer file or to transmit it across an insecure channel such as the Internet. In either scenario the fact that the document is encrypted does not prevent unauthorized people gaining access to it but, rather, ensures that they cannot understand what they see.

The information to be concealed is often called the *plaintext* and the operation of disguising it is known as *encryption*. The encrypted plaintext is called the *ciphertext* or *cryptogram* and the set of rules used to encrypt information plaintext is the *encryption algorithm*. Normally the operation of this algorithm depends on an *encryption key*, which is input to the algorithm together with the message. In order that the recipient can obtain the message from the cryptogram there has to be a *decryption algorithm* which, when used with the appropriate *decryption key*, reproduces the plaintext from the ciphertext.

In general the set of rules that constitute one of these *cryptographic algorithms* is likely to be very complicated and they need careful design. However, for the purpose of this book, the reader may regard them as 'magic formulae' that, with the assistance of keys, transform information into an unreadable form.

The figure provides a diagrammatic description of the use of a *cipher system* to protect a transmitted message.

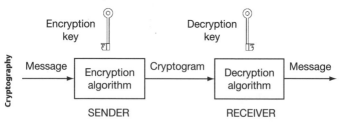

Any person who intercepts a message during transmission is called, not surprisingly, an *interceptor*. Other authors use different terms, including 'eavesdropper', 'enemy', 'adversary', or even 'bad guy'. However, it must be recognized that, on occasions, the interceptors may be the 'good guys'; more about them later. Even if they know the decryption algorithm, interceptors do not, in general, know the decryption key. It is this lack of knowledge that, it is hoped, prevents them from knowing the plaintext. *Cryptography* is the science of designing of cipher systems, whereas *cryptanalysis* is the name given to the process of deducing information about plaintext from the ciphertext without being given the appropriate key. *Cryptology* is the collective term for both cryptography and cryptanalysis.

It is very important to realize that cryptanalysis may not be the only means by which an attacker can gain access to the plaintext.

Suppose, for instance, that someone stores encrypted data on their laptop. Clearly they need to have some way of recovering the decryption key for themselves. If this involves writing it down on a piece of paper which they stick to the lid of the laptop, then anyone who steals the laptop automatically has the decryption key and has no need to perform any cryptanalysis. This is just one simple illustration of the fact that there is certainly more to securing data than using a good encryption algorithm. In fact, as we emphasize repeatedly, the security of the keys is critical for security of cryptographic systems.

In practice most cryptanalytic attacks involve trying to determine the decryption key. If successful, the attacker then has the same knowledge as the intended recipient and is able to decrypt all other communications until the keys are changed. However there may be instances where an attacker's sole objective is to read a particular message. Nevertheless when authors refer to an algorithm as being *broken*, they usually mean that an attacker has found a practical way of determining the decryption key.

Of course, the attacker is only able to break an algorithm if they have sufficient information to enable them to recognize the correct key or, more frequently, to identify incorrect keys. It is important to realize that this extra information is likely to be crucial to the attacker. Suppose, for instance, that they know the plaintext was English text, and that the decryption of some ciphertext using a guessed key does not give meaningful English plaintext. In this case, the guessed key must be incorrect.

One important fact that should already be clear from our introduction is that knowledge of the encryption key is not necessary for obtaining the message from the cryptogram. This simple observation is the basis of the seminal Diffie–Hellman paper. It has had a dramatic impact on modern cryptology and has led to a natural division into two types of cipher systems: symmetric and asymmetric.

A cipher system is called *conventional* or *symmetric* if it easy to deduce the decryption key from the encryption key. In practice, for symmetric systems, these two keys are often identical. For this reason, such systems are frequently called *secret key* or *one-key* systems. However, if it is practically impossible to deduce the decryption key from the encryption key, then the system is called *asymmetric* or *public key*. One reason for distinguishing between these two types of system should be clear. In order to prevent an interceptor with knowledge of the algorithm from obtaining the plaintext from intercepted ciphertext it is essential that the decryption key should be secret. Whereas for a symmetric system this necessitates that the encryption key should also be secret, if the system is asymmetric then knowledge of this key is of no practical use to the attacker. Indeed it can be, and usually is, made public. One consequence of this is that there is no need for the sender and receiver of a cryptogram to share any common secrets. In fact there may be no need for them to trust each other.

Although the statements made in the last paragraph may appear to be simple and self-evident, their consequences are far-reaching. Our diagram above assumes that the sender and recipient have a 'matching pair' of keys. It may, in practice, be quite difficult for them to reach this situation. If, for instance, the system is symmetric then there may be a need to distribute a secret key value before secret messages can be exchanged. The problem of providing adequate protection for these keys should not be underestimated. In fact the general problem of key management, which includes key generation, distribution, storage, change, and destruction, is one of the most difficult aspects of obtaining a secure system. The problems associated with key management tend to be different for symmetric and asymmetric systems. If the system is symmetric then, as we have seen, there may be a need to be able to distribute keys while keeping their values secret. If the system is asymmetric then it may be possible to avoid this particular problem by distributing only the encryption keys, which do not need to be secret. However it is then replaced by the problem of guaranteeing the authenticity of each participant's encryption key, that is, of guaranteeing that the person using a public encryption key value knows the identity of the 'owner' of the corresponding decryption key.

When we were introducing the difference between symmetric and asymmetric systems we assumed that the attacker knew the algorithm. This, of course, is not always true. Nevertheless it is probably best for the designer of a cipher system to assume that any would-be attacker has as much knowledge and general intelligence information as possible. There is a famous principle of cryptography which asserts that the security of a cryptographic system must not depend on keeping secret the cryptographic algorithm. Thus the security should depend only on keeping secret the decryption key.

One of the objectives of studying cryptography is to enable anyone wishing to design or implement a cipher system to assess whether or not that system is secure enough for the particular implementation. In order to assess the security of a system we make the following three assumptions, which we refer to as the *worst-case conditions*.

(**WC1**) The cryptanalyst has a complete knowledge of the cipher system.
(**WC2**) The cryptanalyst has obtained a considerable amount of ciphertext.
(**WC3**) The cryptanalyst knows the plaintext equivalent of a certain amount of the ciphertext.

In any given situation it is necessary to attempt to quantify realistically what is meant by 'considerable' and 'certain'. This depends on the particular system under consideration.

Condition WC1 implies that we believe there should be no reliance on keeping details of the cipher system secret. However this does not imply that the system should be made public. Naturally the attacker's task is considerably harder if he does not know the system used and it is now possible to conceal this information to a certain extent. For instance, with modern electronic systems, the encryption algorithm may be concealed in hardware by the use of microelectronics. In fact it is possible to conceal the entire algorithm within a small 'chip'. To obtain the algorithm an attacker needs to 'open up' one of these chips. This is likely to be a delicate and time-consuming process. Nevertheless it can probably be done, and we should not assume that an attacker lacks the ability and patience to do it. Similarly, any part of the algorithm that is included as software within the machine can be disguised by a carefully written program. Once again, with patience and skill, this can probably be uncovered. It is even possible that, in some situations, the attacker has the precise algorithm available to him. From any manufacturer's or designer's point of view, WC1 is an essential assumption, since it removes a great deal of the ultimate responsibility involved in keeping a system secret.

It should be clear that WC2 is a reasonable assumption. If there is no possibility of interception, then there is no need to use a cipher system. However, if interception is a possibility then, presumably, the communicators are not able to dictate when the interceptions takes place and the safest option is to assume that all transmissions can be intercepted.

WC3 is also a realistic condition. The attacker might gain this type of information by observing traffic and making intelligent guesses. He might also even be able to choose the plaintext for which the ciphertext is known. One 'classic' historical example of this occurred in the Second World War when a light buoy was subjected to a bombing attack merely to ensure that the distinctive German word *Leuchttonne* would appear in plaintext messages that were to be enciphered using the Enigma encryption machine. (See the BBC publication *The Secret War* by B. Johnson.)

An attack which utilizes the existence of known plaintext and ciphertext pairs is called a *known plaintext attack*. If the plaintext is selected by the attacker, as was the situation with the example of bombing light buoys discussed above, then it is a *chosen plaintext attack*. Finally, an attack which has direct knowledge only of the ciphertext is known as a *ciphertext-only* attack.

One consequence of accepting these worst-case conditions is that we have to assume that the only information which distinguishes the genuine recipient from the interceptor is knowledge of the decryption key. Thus the security of the system is totally dependent on the secrecy of the decryption key. This reinforces our earlier assertion about the importance of good key management.

We must stress that assessing the security level of a cipher system is not an exact science. All assessments are based upon assumptions, not only on the knowledge available to an attacker but also on the facilities available to them. The best general principle is, undoubtedly, when in doubt assume the worst and err on the side of caution. It is also worth stressing that, in general, the relevant question is not 'Is this an exceptionally secure system?' but, rather, 'Is this system secure enough for this particular application?' This latter observation is very important and it must be recognized that there is a requirement for cheap, low-level security in certain situations. For almost all non-military implementations, the provision of security is a costly overhead that needs justification from the business perspective. Furthermore the addition of the security facilities may be expensive and frequently degrades the overall performance of the system. Thus there is a natural requirement to keep the security to a minimum. One common way of trying to determine the level of security required is to try to estimate the length of time for which the information needs protection. If we call this the desired *cover time* of the system, then we have a crude indication of the security level required. For instance the cipher system suitable for a tactical network with a cover time of a few minutes may be considerably 'weaker' than that required for a strategic system where, as in the case of government secrets or medical records, the cover time may be decades.

If we assume that our decryption algorithm is known then there is one obvious method of attack available to any adversary. They could, at least in theory, try each possible decryption key and 'hope' that they identify the correct one. Such an attack is called an *exhaustive key search* or, alternatively, a *brute force attack*. Of course such an attack cannot possibly succeed unless the attacker has some way of recognizing the correct key or, as is more common, at least being able to eliminate most incorrect ones. In a known plaintext attack, for instance, it is clear that any choice of the decryption key which does not give the correct plaintext for all the corresponding ciphertext cannot possibly be the correct one. However, as we see when we consider some simple examples, unless there is sufficient volume of corresponding plaintext and ciphertext pairs, there may be many incorrect choices for the decryption key which give the correct answers for all the available ciphertext. If the underlying language of the communications is sufficiently structured, then the statistics of the language can also be used to eliminate certain keys.

We are already in a position where we can begin to give some very basic criteria for assessing the suitability of a given cipher system for any particular application. The users of the system specify a cover time. The designers should know the number of decryption keys. If the designers now make assumptions about the speed with which an attacker could try each key, then they can estimate the expected time for an exhaustive key search to reveal the key. If this latter time is shorter than the cover time, then the system is clearly too weak. Thus our first crude requirement is: the estimated time required for an exhaustive key search should be significantly longer than the cover time.

When distinguishing between symmetric and asymmetric algorithms, we spoke of the need for trust between the sender and receiver. For centuries, prior to the publication of the famous Diffie–Hellman paper, it had been assumed that encrypted messages could only be exchanged between mutually trusting parties. The concept of being able to send messages to someone who was not trusted was considered impossible. Public key algorithms are discussed in later chapters. However, we include here a well-known example of how it is possible for someone to ensure that a present is delivered safely to the intended recipient, despite the fact that it may pass through the possession of many adversaries who would like to possess it.

For the example we assume that the sender has a present, which he wants to lock in a briefcase with a padlock and send to someone, whom he is not prepared to trust with his own key. Instead the sender instructs the intended receiver to buy his own padlock and key. We assume that no one else can find a key that can unlock either the sender's or receiver's padlocks and that the padlocks and briefcase are sufficiently robust that no one can forcibly remove the present from the briefcase. The sender and receiver now carry out the following steps to ensure delivery of the present.

Step 1: The sender places the present in the briefcase, which they lock with their padlock and remove their key. They then send the locked briefcase to the receiver.

Note: While the briefcase is en route from sender to receiver, it is safe from all adversaries, because they cannot remove the padlock from the briefcase. However, the receiver is also unable to obtain the present.

Step 2: The receiver locks the briefcase with their own padlock and removes the key. They then return it to the sender.

Note: The briefcase is now locked with two padlocks so no one can get the present.

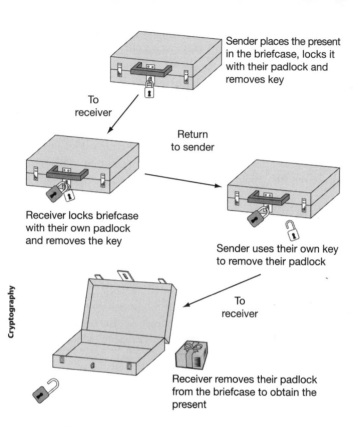

Sender places the present in the briefcase, locks it with their padlock and removes key

To receiver

Return to sender

Receiver locks briefcase with their own padlock and removes the key

Sender uses their own key to remove their padlock

To receiver

Receiver removes their padlock from the briefcase to obtain the present

Step 3: The sender uses their own key to remove their padlock from the briefcase and returns the briefcase to the receiver.

Note: The only lock on the briefcase belongs to the receiver.

Step 4: The receiver removes their padlock from the briefcase to obtain the present.

The result of this sequence of events is that the present has been delivered to the receiver but that neither the receiver nor the sender needed to let their own keys leave their possession. They had no need to trust each other. Of course it is extremely unlikely that our assumptions about the keys and the robustness of the padlocks and briefcase are realistic. However, when we discuss public key cryptography these assumptions are replaced by mathematical equivalents, which are believed to be more acceptable. The point we have illustrated, at least in theory, is that the concept of secure communications without mutual trust might be possible.

In this simplistic example we have to admit that the sender has no way of knowing whose padlock is on the briefcase and that it might be possible for an adversary to impersonate the receiver and place their padlock on the briefcase. This is a problem that has to be addressed. The 'Whose padlock is it?' question in this briefcase example is similar to the 'Whose public key is it?' question that is so important when public key systems are used.

Before we develop the theory further, we include some simple historical examples in the next chapter to illustrate the theory and ensure that the definitions are understood.

Chapter 3

Historical algorithms: simple examples

Introduction

In this chapter we introduce some early 'pencil and paper' examples to illustrate the basic ideas introduced in Chapter 2. We also use them to give some insight into the type of attacks that might be launched by interceptors and illustrate some of the difficulties faced by algorithm designers. The algorithms discussed are all symmetric and were designed, and used, long before public key cryptography was proposed. The chapter is intended for a non-mathematical reader. However, there are a few occasions when we feel it is necessary to discuss the underlying mathematics, especially modular arithmetic. When this occurs the reader's understanding should not be impaired if they skip the mathematics. Nevertheless we provide a basic mathematics tutorial (in the appendix at the end of the chapter) that should be sufficient to enable all readers to understand the text if they wish.

These algorithms are outdated and not really indicative of any modern cryptographic techniques. However, it is very informative to study a number of early systems where encryption was achieved by replacing each letter by another, called letter substitution, and/or changing the order of the letters. There are a number of reasons for this. One is that they enable us to give simple, easy to understand examples that clarify the basic concepts and enable us to illustrate a number of potential weaknesses in ciphers. Another is that they are fun to play with and, since they tend to be non-mathematical, can be enjoyed by 'amateurs' with no scientific training.

Caesar Cipher

One of the earliest examples of a cipher was the *Caesar Cipher* described by Julius Caesar in the *Gallic Wars*. In this cipher each of the letters *A* to *W* is encrypted by being represented by the letter that occurs three places after it in the alphabet. The letters *X, Y, Z* are represented by **A**, **B**, and **C** respectively. Although Caesar used a 'shift' of 3, a similar effect could have been achieved using any number from 1 to 25. In fact any shift is now commonly regarded as defining a Caesar Cipher.

Once again we use a diagram to illustrate a cipher. The figure represents two concentric rings of which the outer one is free to rotate. If we start with the letter **A** outside *A*, a shift of 2 results in **C** being outside *A* and so on. Including the shift of 0 (which of course is the same as a shift of 26), there are now 26 settings. For a Caesar Cipher the encryption key and the decryption key are both determined by a shift.

Once a shift has been agreed then encryption for a Caesar Cipher is achieved by regarding each letter of the plaintext as being on the inside ring and replacing it by the letter outside it in the diagram. For decryption we merely perform the reverse operation. Thus, from the figure, for a shift of 3 the ciphertext for the plaintext message *DOG* is **GRJ** while the plaintext for a cryptogram **FDW** is *CAT*. In order to give the reader confidence that they understand the system we list four statements to check. If the shift is 7 then the ciphertext corresponding to *VERY* is **CLYF** while, for shift 17, the plaintext corresponding to **JLE** is *SUN*.

In our description of the Caesar Cipher, the encryption key and decryption key are both equal to the shift but the encryption and

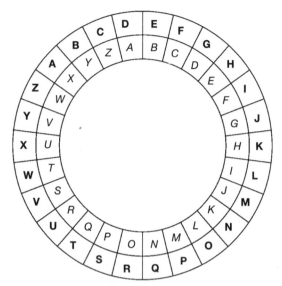

A 'machine' to implement a Caesar Cipher

decryption rules are different. However, we could have changed the formulation slightly to make the two rules coincide and have different encryption and decryption keys. To see this we merely observe that a shift of 26 has the same effect as a shift of 0 and, for any shift from 0 to 25, encryption with that shift is the same as decryption with the new shift obtained by subtracting the original shift from 26. So, for example, encryption with shift 8 is the same as decryption with shift 26 − 8 = 18. This enables us to use the same rule for encryption and decryption with the decryption key 18 corresponding to the encryption key 8.

We have already mentioned exhaustive key searches and clearly, since there are only 26 keys, Caesar Ciphers are vulnerable to this type of attack. Before we give an example we must point out another of their weaknesses. The key can be determined from knowledge of a single pair of corresponding plaintext and ciphertext characters. This is a very small amount of information.

The simplest way to illustrate an exhaustive key search is to work through a complete example that, since there are only 26 keys, is particularly easy for a Caesar Cipher. Let us suppose that we know that a Caesar Cipher is being used, that we anticipate a message in English, and that we intercept the cryptogram **XMZVH**. If the sender had used a shift of 25 for encryption then decryption would be achieved by performing a shift of 1 to give *YNAWI* as the message. Since this has no meaning in English we can safely eliminate 25 as the key value. The result of systematically trying the keys 25 to 1, in decreasing order, is shown in Table 1.

Table 1. An example of an exhaustive key search: cryptogram XMZVH

Encipher-ing key	Assumed 'message'	Encipher-ing key	Assumed 'message'	Encipher-ing key	Assumed 'message'
0	*XMZVH*	17	*GVIEQ*	8	*PERNZ*
25	*YNAWI*	16	*HWJFR*	7	*QFSOA*
24	*ZOBXJ*	15	*IXKGS*	6	*RGTPB*
23	*APCYK*	14	*JYLHT*	5	*SHUQC*
22	*BQDZL*	13	*KZMIU*	4	*TIVRD*
21	*CREAM*	12	*LANJV*	3	*UJWSE*
20	*DSFBN*	11	*MBOKW*	2	*VKXTF*
19	*ETGCO*	10	*NCPLX*	1	*WLYUG*
18	*FUHDP*	9	*ODQMY*		

The only English word among the 26 potential messages is *CREAM* and, consequently, we can deduce that the encryption key is 21. This enables us to decrypt all future messages until the key is changed. Despite the total success of this particular key search, it is important to realize that in general, for more complicated ciphers, a single key search may not identify the key uniquely. It is much more likely merely to limit the number of possibilities by eliminating some obviously wrong ones. As an illustration we revert to the Caesar Cipher and note that an exhaustive search for the encryption key for cryptogram **HSPPW** yields two possibilities that lead to complete English words for the assumed message. (These shifts are 4, that gives *DOLLS*, and 11, that gives *WHEEL*.)

When this happens we need more information, possibly the context of the message or some extra ciphertext, before we can determine the key uniquely. Nevertheless the result of the search does mean that we have significantly reduced the number of possible keys and that, if we do intercept a further cryptogram, we need not conduct another full key search. In fact, for this small illustration, we need only try the values 4 and 11.

There is another interesting observation resulting from this example. During its solution the reader will have found two five-letter English words such that one is obtained from the other by using a Caesar Cipher with shift of 7. You might like to pursue this further and try to find pairs of longer words, or possibly even meaningful phrases, which are shifts of each other.

From this brief illustration it should be clear that Caesar Ciphers are easy to break. Nevertheless they were successfully employed by Julius Caesar, probably because it never occurred to his enemies that he was using a cipher. Today everyone is much more aware of encryption.

The description of a Caesar Cipher using the Caesar Wheel can be replaced by a mathematical definition. We include it here, but anyone who is in any way nervous about the use of mathematical notation should skip this discussion and progress to the next section.

When introducing the Caesar Cipher we noted that a shift of 26 is the same as a shift of 0. This is because a shift of 26 is a complete revolution of the Caesar Wheel. This reasoning can be extended to show that any shift is equivalent to a shift between 0 and 25. For example a shift of 37 is obtained by one complete revolution of the Caesar Wheel and then a shift of 11. Instead of saying, for example, that a shift of 37 is equivalent to a shift of 11, we write $37 = 11 \pmod{26}$.

This is known as using *arithmetic modulo 26*, and 26 is known as the *modulus*. Modular arithmetic, for many other moduli as well as 26, plays a crucial role in a number of cryptographic areas. Thus at the end of this chapter we include an appendix to familiarize the reader with the relevant definitions and results in this branch of elementary number theory.

Caesar Ciphers are sometimes referred to as *additive ciphers*. In order to appreciate why, we have only to assign integer values to the letters in the following way:

$$A = 0, B = 1, \ldots, Z = 25.$$

Encryption for a Caesar Cipher with shift y can now be achieved by replacing the number x by $x + y \pmod{26}$. Thus, for example, since N is the fourteenth letter of the alphabet, $N = 13$. To encrypt N with a shift of 15, we have $x = 13$, $y = 15$, which means the encrypted version of N is $13 + 15 = 28 = 2 \pmod{26}$. So N is encrypted as **C**.

As we have already seen, the number of keys for additive ciphers is too small. If we try to think of ways of obtaining a cipher system with a larger number of keys then we might try extending the system to include multiplication as an alternative encryption rule. However if we try this then, since encryption must be a reversible process, there is restriction on the number of 'multiplicative keys'.

Suppose we try to encrypt messages by multiplying by 2 and use arithmetic modulo 26. When we do this A and N are both encrypted as **A**, B and O are both encrypted as **C**, etc. It turns out that only those letters represented by even numbers can appear in any cryptogram and that each such cryptogram letter might represent one of two letters. This makes decryption virtually impossible and multiplication by 2 cannot be used for encryption. An even more dramatic example is trying to encrypt by multiplying by 13. In this case half the alphabet would be encrypted as **A** with the other half as **N**. In fact 1, 3, 5, 7, 9, 11, 15, 17, 19, 21, 23, or 25 are the only numbers for which multiplication can be used to provide encryption.

Simple Substitution Ciphers

Although having a large number of keys is a necessary requirement for cryptographic security, it is important to stress that having a large number of keys is certainly no guarantee that the cipher system is strong. A common example is the Simple Substitution Cipher (or Monoalphabetic Cipher) that we now discuss in detail. The discussion of this cipher not only establishes the dangers of relying on a large number of keys as an indication of strength, but also illustrates how the statistics of the underlying language, in this case English, may be exploited by an attacker.

For a Simple Substitution Cipher we write the alphabet in a randomly chosen order underneath the alphabet written in strict alphabetical order. An example is given here.

A	B	C	D	E	F	G	H	I	J	K	L	M
D	**I**	**Q**	**M**	**T**	**B**	**Z**	**S**	**Y**	**K**	**V**	**O**	**F**

N	O	P	Q	R	S	T	U	V	W	X	Y	Z
E	**R**	**J**	**A**	**U**	**W**	**P**	**X**	**H**	**L**	**C**	**N**	**G**

The encryption and decryption keys are equal. They are simply the order in which the bold letters are written. The encryption rule is 'replace each letter by the one beneath it' while the decryption rule is the opposite procedure. Thus, for example, for the key in this figure, the cryptogram corresponding to *GET* is **ZTP**, while the message corresponding to **IYZ** is *BIG*. Note, by the way, that the Caesar Cipher is a special case of a Simple Substitution Cipher where the order in which the bold letters are written is merely a shift of the alphabet.

The number of keys for a Simple Substitution Cipher is equal to the number of ways in which the 26 letters of the alphabet can be arranged. It is called 26 factorial and is denoted by 26! It is $26 \times 25 \times 24 \times \ldots \times 3 \times 2 \times 1$ which equals

403,291,461,126,605,635,584,000,000.

This is undoubtedly a large number and no one is likely to attempt to discover the key by an exhaustive search. However having such a large number of keys introduces its own problems, and there are a number of observations to be made about the key management problems associated with the use of Simple Substitution Ciphers. The first obvious comment is that, unlike the Caesar Cipher, the key is long and difficult to memorize. Thus when, in pre-computer days, this type of system was used manually, the key was frequently written on a piece of paper. If this paper was seen and/or stolen then the system was compromised. If the paper was lost then all encrypted messages were 'lost' in the sense that the genuine recipient had to break the algorithm to determine them.

In order to overcome this type of danger, users tried to find methods of generating keys that were easy to remember. One common method was to think of a key phrase, remove all repetitions of each letter, let this form the 'beginning' of the key and then extend it to a full key by adding the remaining letters in alphabetical order. So, for instance, if we let our key phrase be 'We hope you enjoy this book' then removing repetitions gives 'wehopyunjtisbk' and thus the key is

W E H O P Y U N J T I S B K A C D F G L M Q R V X Z

Clearly restricting keys to those that can be derived from key phrases reduces the number of keys because a significant proportion of the 26! possible simple substitution keys cannot be derived from an English phrase in this way. However, the number of keys is still too large for an exhaustive search to be feasible and it is now easy to remember keys.

A second obvious observation about Simple Substitution Ciphers is that it is quite likely that many different keys encrypt the same message into the same cryptogram. Suppose, for example, that the message is *MEET ME TONIGHT*. If we use our first example of a key then the cryptogram is **FTTP FT PREYZSP**. However any key which maps *E* to **T**, *G* to **Z**, *H* to **S**, *I* to **Y**, *M* to **F**, *N* to **E**, *O* to **R** and *T* to **P** also produces that same cryptogram. There are

$$18! = 6{,}402{,}373{,}705{,}728{,}000$$

such keys. Clearly this means that, at least for this type of cipher, we should not assume that the attacker needs to determine the complete key before obtaining our message from an intercepted cryptogram.

Before we discuss how the statistics of the English language can be exploited to attack a number of ciphers, including the Simple Substitutions, we illustrate some particular properties of Simple Substitution Ciphers by considering four small, carefully chosen examples. In the following examples we assume that the cryptograms given have been intercepted by someone who knows that the message is in English and that a Simple Substitution Cipher was used.

Example 1: G WR W RWL

Since there are only two one-letter words in English it is reasonable to assume that either **G** represents *A* and **W** represents *I* or vice versa. It is then easy to eliminate the possibility that **G** is *A* and we quickly come to the conclusion that the message begins *I AM A MA* and that there are a limited number of possibilities for the last letter. If we happen to know that the message is a complete sentence in English then it is almost certain to be *I AM A MAN*. It is important to realize that this simple reasoning does not use any cryptanalytic techniques. It is more or less 'forced' by the structure of the English language. Note also that, although it definitely does not determine the key, it reduces the number of possible keys from 26! to 22!. If this were the beginning of a longer message then we would either need other arguments to determine the rest of the key or need to perform a reduced, but still computationally infeasible, key search. We note also that it was common practice to prevent this type of attack by transmitting the letters in groups of five, thereby removing all information about word sizes and/or word endings.

Example 2: HKC

What can we say? Not much. Since there is no other information, the message could be any meaningful sequence of three distinct letters. We can almost certainly eliminate a few keys, for example, those that simultaneously encrypt *Z* as **H**, *Q* as **K**, and *K* as **C**. However the number of remaining possibilities is so high that we might also be tempted to say that, on its own, intercepting this cryptogram tells us nothing. It is certainly true that if we wanted to send just one three-letter message then a Simple Substitution Cipher would appear adequate and that an exhaustive key search for the cryptogram would yield all three-letter words (with distinct letters) as potential messages.

Example 3: HATTPT

For this example we can certainly limit the number of possibilities for the number of plaintext letters that might be mapped onto **T**. We might also safely deduce that one of **T** or **P** represents a vowel. Furthermore, if we had reason to believe that the intercepted message was a complete, single word then we might be able to write down all possibilities. A few examples are *CHEESE*, *MISSES*, and *CANNON*.

Example 4: HATTPT (Given that the message is the name of a country)

For this example we believe that the message has to be *GREECE*. The difference between Examples 3 and 4 is that for Example 4 we have some extra information that transforms the attacker's task from being 'impossible' to being easy. This, of course, is one of the functions of intelligence departments in 'war situations'. It is often their intelligence information that is the deciding factor that enables cryptanalysts to break the enemy's ciphers.

The statistics of the English language

The examples in the last section were all short and were carefully chosen to illustrate specific points. However, even if a Simple Substitution Cipher is used to encrypt reasonably long passages of English text, there are a number of straightforward attacks which reveal the message and the key, or at least the major part of the key. These attacks exploit some well-known characteristics of the English language. Table 2 shows the frequencies, expressed as a percentage, of the letters of the alphabet in a sample of over 300,000 characters taken from passages from numerous newspapers and novels. (This table is based on one that was originally published in *Cipher Systems: The Protection of Communications* and was compiled by H. J. Beker and F. C. Piper.)

They are consistent with numerous other tables compiled by other authors and can be interpreted as presenting the expected frequencies of the letters in English text. They show, quite clearly, that English text is likely to be dominated by a very small number of letters.

When a Simple Substitution Cipher is used each specific letter of the alphabet is replaced by the same substituted letter, no matter where it appears in the text. Thus if, for example, we use a cipher where **R** replaces *E*, then the frequency of **R** in the cryptogram is equal to the frequency of *E* in the message. This means that if the

Table 2. The expected relative frequencies of letters in English text

Letter	%	Letter	%
A	8.2	N	6.7
B	1.5	O	7.5
C	2.8	P	1.9
D	4.2	Q	0.1
E	12.7	R	6.0
F	2.2	S	6.3
G	2.0	T	9.0
H	6.1	U	2.8
I	7.0	V	1.0
J	0.1	W	2.4
K	0.8	X	2.0
L	4.0	Y	0.1
M	2.4	Z	0.1

frequency of the letters in a message is reflected by Table 2 then the frequencies for the cryptogram show the same imbalance but with the frequencies distributed differently amongst the letters. To see this more clearly we show the frequency histogram for a long cryptogram that was obtained using a Simple Substitution Cipher.

By comparing Table 2 with this figure, a cryptanalyst might reasonably guess that **H** corresponds to *E* and that **W** corresponds to *T*. Since the most popular trigram in the English language is overwhelmingly *THE*, the attacker can then gain confidence in this assumption by seeing if the most popular trigram in the cryptogram is **W*H**, where * represents a fixed letter. This not only confirms the original guesses but also suggest that the plaintext equivalent of the letter * is *H*. Anyone interested in seeing how easy it is to break these ciphers should try to read the next paragraph that has been encrypted using a Simple Substitution Cipher.

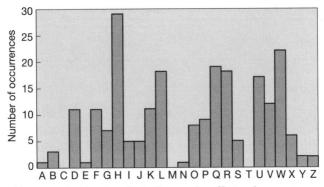

A histogram showing the relative frequencies of letters in a cryptogram which was obtained using a Simple Substitution Cipher

DIX DR TZX KXCQDIQ RDK XIHPSZXKPIB TZPQ TXGT PQ TD QZDM TZX KXCJXK ZDM XCQPVN TZPQ TNSX DR HPSZXK HCI LX LKDUXI. TZX MDKJ QTKFHTFKX DR TZX SVCPITXGT ZCQ LXXI SKXQXKWXJ TD OCUX TZX XGXKHPQX XCQPXK. PR MX ZCJ MKPTTXI TZX. HKNSTDBKCOPI BKDFSQ DR RPWX VXTTXKQ TZXI PT MDFVJ ZCWX LXXI ZCKJXK. TD HDIWPIHX NDFKQXVWXQ DR TZPQ SCPKQ SCPKQ DR KXCJXKQ HCI SKDWPJX XCHZ DTZXK MPTZ HKNSTDBKCOQ MPTZ TZPQ VXTTXK BKDFSIB.

Any reader who has decrypted this cryptogram will almost certainly have made use of the information given by the spacing. The problem would have been significantly harder had the original English language spacing been removed.

We end this brief discussion by confessing that we have not quantified our concept of a 'long' cryptogram. There is, of course, no precise answer. While 200 characters are almost certainly sufficient for the statistics to be reliable, we have found that students can usually determine the message for cryptograms of about 100 or more characters.

As an aside, we stress that there is no guarantee that the statistics of any particular message agrees precisely with Table 2. If, for instance, a personal letter is being encrypted then it is quite likely that the word 'you' is almost as popular as 'the'. As an illustration of how the statistics of a message might be deliberately manipulated, we note that there is a 200-page novel that does not use the letter E (Gilbert Adair's translation of *A Void* by Georges Perec).

The reason that the type of attack just illustrated is possible is that there are a few 'popular' letters that are likely to 'dominate' the message and their ciphertext equivalents can then readily be identified. One way to prevent this might be to perform a simple substitution on the *bigrams*, that is, pairs of consecutive letters. If we did this then our key would be an arrangement of the 676 bigrams. This would give us very long keys and an astronomically large key space of 676! keys. However it would be very clumsy and would also be subject to the same type of attack since, just as for single letters, long messages are likely to be dominated by comparatively few bigrams.

Clearly it would not be practical to try to present a list of all 676 bigrams with their cryptogram equivalent beneath them, that is, to mimic the representation of our original Simple Substitution Cipher key. Consequently we need some easy way of determining keys and of expressing the encryption and decryption algorithm. We now give an example of a cipher that operates on bigrams but uses only comparatively few of all possible keys.

The Playfair Cipher

The *Playfair Cipher* was invented by Sir Charles Wheatstone and Baron Lyon Playfair in 1854 and was used by the British War Office up to the beginning of the 20th century, including use in the Boer War. It is an example of a 'Bigram' Cipher. This means that letters are encrypted in pairs, as opposed to individually. The key is a 5 by 5 square (with entries comprised of the 25 letters obtained by deleting

J from the alphabet) and thus there are 25! or

$$15{,}511{,}210{,}043{,}330{,}985{,}984{,}000{,}000$$

keys. Before encrypting using a Playfair Cipher the message has to be rearranged slightly. To do this you:

- replace Js with Is;
- write message in pairs of letters;
- do not allow identical pairs – if they occur insert Z between them;
- if the number of letters is odd, add Z to the end.

In order to illustrate how the cipher works we choose a specific key. However, there is nothing special about our choice.

S	T	A	N	D
E	R	C	H	B
K	F	G	I	L
M	O	P	Q	U
V	W	X	Y	Z

Once the message has been suitably rearranged we give the rule for encryption. In order to clarify our description we extend the key by adding a sixth column and a sixth row to the original key. The sixth row is identical to the first row, while the sixth column is identical to the first column. Thus, for our example, the extended key can be set out as in the diagram.

S	T	A	N	D	S
E	R	C	H	B	E
K	F	G	I	L	K
M	O	P	Q	U	M
V	W	X	Y	Z	V
S	T	A	N	D	

32

The rule for encryption is as follows.

- If the two letters lie in the same row of the key then each letter is replaced by the one on its right in the extended key.
- If two letters lie in the same column of the key then each letter is replaced by the one below it in the extended key.
- If the two letters are not in the same row or column then the first letter is replaced by the letter that is in the row of the first letter and the column of the second letter. The second letter is replaced by the fourth corner of the rectangle formed by the three letters used so far.

We now encrypt the message: *GOOD BROOMS SWEEP CLEAN*

Since there are no Js in the message we have only to write the message in pairs of letters with the appropriate insertion of extra Zs. This gives:

GO   OD   BR   OZ   OM   SZ   SW   EZ   EP   CL   EA   NZ

Thus, for our chosen key, *GO* becomes **FP**; *OD* becomes **UT**; *OM* becomes **PO**. The complete cryptogram is

FP   UT   EC   UW   PO   DV   TV   BV   CM   BG   CS   DY

As with Simple Substitution Ciphers, users tended to use a key phrase to determine the key matrix. The method was the same as that used for Simple Substitution Ciphers, that is, write out the key phrase, remove repeated letters and then add the unused letters in alphabetical order. So, if the key phrase is UNIVERSITY OF LONDON then removing repeated letters gives UNIVERSTYOFLD and the square can be set out as in the next diagram.

Decryption is, as always, merely the reverse process to encryption. Any reader wishing to ensure they understand how the Playfair Cipher works should decrypt **MBOUBTZE** using the square below as key. (The answer is a seven-letter English word that, we hope,

U	N	I	V	E
R	S	T	Y	O
F	L	D	A	B
C	G	H	K	M
P	Q	W	X	Z

does not reflect the reader's state of mind.) We do not intend to discuss the cryptanalysis of this cipher. There are many other examples of ciphers that are easy to describe and fun to play with. Suitable references are provided at the end of this book.

Homophonic Coding

Another option for improving on the Simple Substitution Cipher might be to expand the alphabet by introducing some extra characters so that, for instance, the plaintext letter E is represented by more than one ciphertext character.

These extra characters are known as randomizing elements and the process of expanding the alphabet is called homophonic coding. As an illustration we might introduce a cipher where the alphabet for the cryptogram would be the numbers 00, 01, 02, ... , 31. Each cryptogram number represents a unique plaintext letter but the letters A, E, N, O, R, T are each represented by two distinct letters.

As an illustration we might assign numbers to letters as in the diagram.

A	A	B	C	D	E	E	F	G	H	I	J	K	L	M	N
01	07	14	21	04	13	17	20	29	31	06	28	12	30	17	00

N	O	O	P	Q	R	R	S	T	T	U	V	W	X	Y	Z
18	26	19	09	10	25	23	02	08	24	22	05	16	15	11	03

If we do this then the word *TEETH*, which has two pairs of repeated letters, might be written as 24 27 13 08 31. To anyone who does not know the key, the five characters of the cryptogram are different but the genuine receiver is in no danger of confusion.

The six letters chosen are likely to be the six most popular letters in the plaintext. If, for instance, the decision about which of the two chosen numbers represents a particular incidence of the letter *E* is random, then we would expect each of the two numbers to 'occupy' about 6 per cent of the cryptogram. In general the effect of homophonic coding is to ensure that the expected frequency histogram of the cryptogram is flatter than that of the plaintext. This makes the statistical attack more difficult.

Note 1: In this cipher we write 00, 01, 02 for 0, 1, 2 etc. Whenever spaces are not used this type of notation is needed to distinguish between, for example, 'twelve' and 'one followed by two'.

Note 2: Breaking ordinary Simple Substitution Ciphers is relatively easy and we hope all readers managed to break the paragraph-long cryptogram above. Breaking the type of cipher we are now discussing requires considerably more patience and luck. Anyone who needs convincing or enjoys this type of puzzle should try to read the following cryptogram. The only information is that English text has been encrypted using a Simple Substitution Cipher with homophonic coding as discussed above. The key is unknown, but is not the one listed above. Furthermore the letters have been written in groups of five. (This means the attacker cannot identify short words, notably those of one letter.) It is not easy and readers should not feel obliged to try.

24 29 25 00 20	01 12 27 10 01	12 06 29 07 08
31 29 05 07 14	20 26 01 04 26	20 06 28 29 28
05 04 31 28 18	30 01 31 21 26	25 24 26 12 29
04 26 31 18 23	15 21 25 26 31	28 26 30 10 01
21 07 31 18 16	12 12 28 18 13	05 08 21 24 30

```
20 21 25 24 21      30 10 18 17 19      31 28 18 05 12
31 05 24 09 21      08 26 05 08 14      12 17 27 07 04
18 20 08 12 05      25 04 13 27 31      12 28 18 19 05
24 31 12 28 05      12 12 28 18 08      31 01 12 21 08
31 21 24 08 05      23 18 19 10 01      12 12 26 23 15
26 05 25 08 21      31 21 08 07 29      12 08 29 26 05
08 14 12 17 21      04 26 25 12 21      19 14 31 28 18
30 17 30 27 10      01 20 10 26 31      12 26 20 08 21
25 12 28 18 30      10 05 21 07 12      18 16 31 30 01
12 21 18 25 24      26 01 07 04 10      27 24 09 05 23
26 13 29 31 28      11 18 20 14 21      15 30 29 20 12
01 07 31 19 17      23 12 28 26 24      23 14 30 12 01
07 01 10 14 08      12 21 25 19 01      24 31 13 20 18
05 09 21 07 00      24 21 30 28 26      20 08 27 08 27
05 10 10 14 21      07 11 29 10 11      18 08 01 15 21
16 31 27 23 26      17 19 08 24 21      18 25 12 21 19
21 24 20 18 01      08 17 07 21 25      00 05 25 04 21
07 08 30 21 20      18 04 00 27 26      08 08 06 17 23
09 21 07 12 28      21 08 24 17 25      31 18 16 31 06
26 25 17 12 18      31 28 01 12 31      28 26 24 20 14
30 12 17 00 20      01 30 28 21 24      12 18 05 15 18
15 30 10 29 14      18 04 01 31 13      10 26 12 24 28
10 26 14 30 05      23 09 21 07 24      10 27 04 26 04
30 26 17 30 10      26 06 21 12 28      05 07 01 30 31
21 31 27 04 18      19 17 23 24 20      17 08 08 06 17
20 04 30 27 03      03 10 26 08
```

Polyalphabetic Ciphers

When a homophonic cipher is used, the frequency histogram of the
cryptogram is made flatter by increasing the size of the alphabet.
This ensures that more than one ciphertext character may represent
the same plaintext character. However, it is still true that each
ciphertext character represents a unique plaintext character and
there is always the danger of an attacker compiling a dictionary of
known plaintext and ciphertext pairs for a given key.

Another way of achieving the objective of flattening the frequency histogram is by the use of a polyalphabetic cipher. When a polyalphabetic cipher is used, the ciphertext character replacing a particular plaintext letter may vary through the cryptogram and might, for instance, depend on its position in the plaintext message or the content of the plaintext that precedes it. For these ciphers it is now true that the same ciphertext character may represent different plaintext letters. This is not true for homophonic coding.

Once again we must point out that the simple examples of these ciphers that we describe are no longer used. However we discuss them in some detail as we can then illustrate a number of weaknesses that the modern algorithm designer has to avoid. As with our earlier examples, we include them to illustrate a number of cryptanalytic techniques and because they enable us to set exercises which are both informative and fun.

Vigenère Ciphers

The *Vigenère Cipher* is probably the best known of the 'manual' polyalphabetic ciphers and is named after Blaise de Vigenère, a 16th-century French diplomat. Although it was published in 1586, it was not widely recognized until nearly 200 years later and was finally broken by Babbage and Kasiski in the middle of the 19th century. It is interesting to note that the Vigenère Cipher was used by the Confederate Army in the American Civil War. The Civil War occurred after the cipher had been broken. This is illustrated by the quotation by General Ulysses S. Grant: 'It would sometimes take too long to make translations of intercepted dispatches for us to receive any benefit from them, but sometimes they gave useful information.'

The Vigenère Cipher uses a Vigenère Square to perform encryption. The left-hand (key) column of this square contains the English alphabet and, for each letter, the row determined by that letter contains a rotation of the alphabet with that letter as the leading

Key | **Plaintext**

	A	B	C	D	E	F	G	H	I	J	K	L	M	N	O	P	Q	R	S	T	U	V	W	X	Y	Z
a	A	B	C	D	E	F	G	H	I	J	K	L	M	N	O	P	Q	R	S	T	U	V	W	X	Y	Z
b	B	C	D	E	F	G	H	I	J	K	L	M	N	O	P	Q	R	S	T	U	V	W	X	Y	Z	A
c	C	D	E	F	G	H	I	J	K	L	M	N	O	P	Q	R	S	T	U	V	W	X	Y	Z	A	B
d	D	E	F	G	H	I	J	K	L	M	N	O	P	Q	R	S	T	U	V	W	X	Y	Z	A	B	C
e	E	F	G	H	I	J	K	L	M	N	O	P	Q	R	S	T	U	V	W	X	Y	Z	A	B	C	D
f	F	G	H	I	J	K	L	M	N	O	P	Q	R	S	T	U	V	W	X	Y	Z	A	B	C	D	E
g	G	H	I	J	K	L	M	N	O	P	Q	R	S	T	U	V	W	X	Y	Z	A	B	C	D	E	F
h	H	I	J	K	L	M	N	O	P	Q	R	S	T	U	V	W	X	Y	Z	A	B	C	D	E	F	G
i	I	J	K	L	M	N	O	P	Q	R	S	T	U	V	W	X	Y	Z	A	B	C	D	E	F	G	H
j	J	K	L	M	N	O	P	Q	R	S	T	U	V	W	X	Y	Z	A	B	C	D	E	F	G	H	I
k	K	L	M	N	O	P	Q	R	S	T	U	V	W	X	Y	Z	A	B	C	D	E	F	G	H	I	J
l	L	M	N	O	P	Q	R	S	T	U	V	W	X	Y	Z	A	B	C	D	E	F	G	H	I	J	K
m	M	N	O	P	Q	R	S	T	U	V	W	X	Y	Z	A	B	C	D	E	F	G	H	I	J	K	L
n	N	O	P	Q	R	S	T	U	V	W	X	Y	Z	A	B	C	D	E	F	G	H	I	J	K	L	M
o	O	P	Q	R	S	T	U	V	W	X	Y	Z	A	B	C	D	E	F	G	H	I	J	K	L	M	N
p	P	Q	R	S	T	U	V	W	X	Y	Z	A	B	C	D	E	F	G	H	I	J	K	L	M	N	O
q	Q	R	S	T	U	V	W	X	Y	Z	A	B	C	D	E	F	G	H	I	J	K	L	M	N	O	P
r	R	S	T	U	V	W	X	Y	Z	A	B	C	D	E	F	G	H	I	J	K	L	M	N	O	P	Q
s	S	T	U	V	W	X	Y	Z	A	B	C	D	E	F	G	H	I	J	K	L	M	N	O	P	Q	R
t	T	U	V	W	X	Y	Z	A	B	C	D	E	F	G	H	I	J	K	L	M	N	O	P	Q	R	S
u	U	V	W	X	Y	Z	A	B	C	D	E	F	G	H	I	J	K	L	M	N	O	P	Q	R	S	T
v	V	W	X	Y	Z	A	B	C	D	E	F	G	H	I	J	K	L	M	N	O	P	Q	R	S	T	U
w	W	X	Y	Z	A	B	C	D	E	F	G	H	I	J	K	L	M	N	O	P	Q	R	S	T	U	V
x	X	Y	Z	A	B	C	D	E	F	G	H	I	J	K	L	M	N	O	P	Q	R	S	T	U	V	W
y	Y	Z	A	B	C	D	E	F	G	H	I	J	K	L	M	N	O	P	Q	R	S	T	U	V	W	X
z	Z	A	B	C	D	E	F	G	H	I	J	K	L	M	N	O	P	Q	R	S	T	U	V	W	X	Y

The Vigenère Square

character. So each letter in the left-hand column gives a Caesar Cipher whose shift is determined by that letter. Thus, for example, the letter g gives the Caesar Cipher with shift 6.

One of the most common methods for using the square to obtain a cipher involves choosing a keyword (or key phrase) with no repeated letters. If the plaintext message is longer than the keyword, then, by repeating the key as often as is necessary, we obtain a sequence of letters which is as long as the message. We then write this sequence of letters beneath our message. Thus, for example, if our message is *PLAINTEXT* and our keyword is 'fred' we obtain:

Message	*P*	*L*	*A*	*I*	*N*	*T*	*E*	*X*	*T*
Key	**f**	**r**	**e**	**d**	**f**	**r**	**e**	**d**	**f**

We now use the square to encrypt the message as follows.

To encrypt the initial letter *P* we use the key letter beneath it which, in this case, is **f**. Thus to encrypt *P* we go to the row of the square determined by **f** and read off the letter beneath *P*, which is **U**. Similarly we encrypt *L* by taking the letter underneath it in the row determined by **r**, which is **C**. The process for encrypting *P* with key letter **f** is illustrated here.

	Key	**Plaintext**

Key		A B C D E F G H I J K L M N O P Q R S T U V W X Y Z
	a	A B C D E F G H I J K L M N O P Q R S T U V W X Y Z
	b	B C D E F G H I J K L M N O P Q R S T U V W X Y Z A
	c	C D E F G H I J K L M N O P Q R S T U V W X Y Z A B
	d	D E F G H I J K L M N O P Q R S T U V W X Y Z A B C
	e	E F G H I J K L M N O P Q R S T U V W X Y Z A B C D
	f	F G H I J K L M N O P Q R S T U V W X Y Z A B C D E
	g	G H I J K L M N O P Q R S T U V W X Y Z A B C D E F
	h	H I J K L M N O P Q R S T U V W X Y Z A B C D E F G
	i	I J K L M N O P Q R S T U V W X Y Z A B C D E F G H
	j	J K L M N O P Q R S T U V W X Y Z A B C D E F G H I
	k	K L M N O P Q R S T U V W X Y Z A B C D E F G H I J
	l	L M N O P Q R S T U V W X Y Z A B C D E F G H I J K
	m	M N O P Q R S T U V W X Y Z A B C D E F G H I J K L
	n	N O P Q R S T U V W X Y Z A B C D E F G H I J K L M
Cryptography	o	O P Q R S T U V W X Y Z A B C D E F G H I J K L M N
	p	P Q R S T U V W X Y Z A B C D E F G H I J K L M N O
	q	Q R S T U V W X Y Z A B C D E F G H I J K L M N O P
	r	R S T U V W X Y Z A B C D E F G H I J K L M N O P Q
	s	S T U V W X Y Z A B C D E F G H I J K L M N O P Q R
	t	T U V W X Y Z A B C D E F G H I J K L M N O P Q R S
	u	U V W X Y Z A B C D E F G H I J K L M N O P Q R S T
	v	V W X Y Z A B C D E F G H I J K L M N O P Q R S T U
	w	W X Y Z A B C D E F G H I J K L M N O P Q R S T U V
	x	X Y Z A B C D E F G H I J K L M N O P Q R S T U V W
	y	Y Z A B C D E F G H I J K L M N O P Q R S T U V W X
	z	Z A B C D E F G H I J K L M N O P Q R S T U V W X Y

Using the Vigenère Square to encrypt *P* with the key letter f

Any reader who completes this process should conclude that the complete cryptogram for *PLAINTEXT* with keyword **fred** is **UCELSLIAY**.

This means we now have:

message:	*P*	*L*	*A*	*I*	*N*	*T*	*E*	*X*	*T*
key:	**f**	**r**	**e**	**d**	**f**	**r**	**e**	**d**	**f**
ciphertext:	**U**	**C**	**E**	**L**	**S**	**L**	**I**	**A**	**Y**

We can now observe that the plaintext letter *T* is represented by both **L** and **Y** in the ciphertext and that the ciphertext letter **L** represents both *I* and *T*. Thus it is clear that, by using these ciphers, we can prevent the frequency of letters in the cryptogram having the same distinctive patterns as those exhibited by cryptograms resulting from Simple Substitution Ciphers.

There are many variations of the Vigenère Cipher, including one in which the keyword is allowed to have repeated letters. Each of them has slightly different characteristics that lead to slightly varying attacks. However we restrict our attention to the precise system that we have defined.

Vigenère Ciphers are a particular example of a polyalphabetic cipher in which a (short) sequence of Simple Substitution Ciphers is repeatedly used in strict rotation. The number of component ciphers used is called the *period* and, clearly, for the version of the Vigenère Cipher that we have described, the period is equal to the length of the keyword.

Before continuing our discussion of periodic ciphers it is worth noting that, for instance, a polyalphabetic cipher of period 3 is nothing more than a particular example of a Simple Substitution Cipher on trigrams. This simple observation is merely a special instance of the general principle that changing the alphabet can change the 'nature' of a cipher. For the present we are concentrating on ciphers where the basic symbols are the letters of the English alphabet. When discussing more modern ciphers, we tend to regard all messages as sequences of binary digits (0s and 1s).

As we have already noted, one motivation for using polyalphabetic ciphers is to disguise the letter frequency of the underlying language. As an illustration of how this is achieved we include a histogram to show the frequency count for a specific cryptogram that was the result of using a Vigenère Cipher of period 3 on a passage of English text.

There are a number of obvious differences between this histogram and the one shown earlier. The most striking ones are that every letter occurs in the second histogram and that no letter dominates this histogram in quite the same way that **H** dominates the first one. This second histogram is undoubtedly flatter than the earlier one and, consequently, is not of such immediate assistance to a would-be attacker. Anyone looking at the second histogram would be tempted to deduce that the cryptogram letter **R** represents the plaintext *E* somewhere. However they would not know at which positions this occurred.

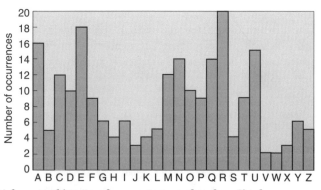

A frequency histogram for a cryptogram when three Simple Substitution Ciphers were used in strict rotation

In general we would expect the flatness of the histogram to reflect the length of the period and that longer periods make the ciphers harder to break. In some sense this is true. However all that using a periodic polyalphabetic cipher achieves in practice is to ensure that the cryptanalyst needs considerably more ciphertext to start an attack. In order to illustrate this, we concentrate on Vigenère Ciphers. Some of our assertions are true for any polyalphabetic cipher, whilst others rely on the characteristics of our definition of a Vigenère Cipher. It is important that the reader distinguishes between the two situations. Thus changing the polyalphabetic cipher may change the details of the attack and 'strengthen' the system slightly. However, polyalphabetic ciphers where the key is significantly shorter than the message are all vulnerable to some variation of the attacks exhibited here.

In order to break a Vigenère Cipher it is sufficient to determine the keyword. If the period is known and is not too long then this can be determined by writing a computer program to conduct an exhaustive key search. As an illustration readers might like to perform a key search on the cryptogram **TGCSZ GEUAA EFWGQ AHQMC**, given that it is the result of using a Vigenère Cipher with a keyword of length 3 on a passage of English text. Any reader who tries this is presented with an interesting problem of recognizing the correct keyword. The basic assumption is, presumably, that it is the only three-letter word which gives plaintext that is meaningful. However the real problem is how can one recognize that the plaintext is meaningful. One possibility is to sit at the screen and inspect the result of applying each key. Clearly this is both boring and time consuming. Alternatives must be found.

When conducting an exhaustive search for a keyword of length p, it is probably easier to try all sequences of p letters systematically than to consider only English words. Thus for a Vigenère Cipher of known period p, an exhaustive key search probably requires 26^p trials. This means that, as the period increases, exhaustive key searches can quickly become unmanageable. However, if the period is known then it is comparatively straightforward to determine the keyword without conducting a search. One way is to write the ciphertext in p rows in such a way that the cryptogram is reconstructed by writing each column in order. So, for instance, for $p = 3$, if the ciphertext is $c_1c_2c_3c_4c_5c_6c_7c_8c_9 \ldots$ then we write it as

$$c_1c_4c_7c_{10} \cdots$$
$$c_2c_5c_8c_{11} \cdots$$
$$c_3c_6c_9c_{12} \cdots$$

Once this is done then each row is the result of using the same Simple Substitution Cipher that, for the particular case of a Vigenère Cipher, is an additive cipher. We can now use the statistical arguments of the last section on each individual row. In fact, for a Vigenère Cipher where the length of the cryptogram is long in comparison to the period p, it is probably sufficient to determine the most frequent letter in each row and assume that it represents either E, T, or A. This latter observation exploits the fact that, for each row, the Simple Substitution Cipher being used is a Caesar Cipher. This means, as we have already noted, that knowledge of a single plaintext and ciphertext pair is sufficient to determine the key. Consequently if the ciphertext equivalent of one single letter for each row can be determined, probably by a suitable mixture of intelligent guessing and luck, then the key can be determined.

The discussion so far suggests that the real problem facing the attacker of a Vigenère Cipher is the determination of the period p. One possibility is to systematically try all small values of p. However, there are also a number of simple, ingenious ways of achieving it. The most famous is known as the Kasiski test and is the one used by Babbage, who was the first person to break the cipher. His method was to search for repeated (long) strings of characters in the ciphertext. When they occur they are quite likely to represent identical passages of the message encrypted using identical keyboard letters. This implies that the gaps between these repeated patterns are likely to be multiples of the period. (The cryptanalysis of the Vigenère Cipher is discussed in detail in Singh's *The Code Book*.)

Transposition Ciphers

So far all the examples given have taken the message and substituted letters, or blocks of letters, with other letters or blocks. Thus they all come under the general heading of Substitution Ciphers. There are, however, other families of ciphers which are based on the idea of transposing the order in which the letters are written. These are known as *Transposition Ciphers*. We give a very simple example.

In our example the key is a small number. We use 5 as the key. In order to encrypt a message using this key, we write the key in rows of 5 letters and encrypt by writing the letters of the first column first, then the second column etc. If the length of the message is not a multiple of 5 then we add the appropriate number of *Z*s at the end before we encrypt. The process is most easily understood by working through a small example.

We encrypt the message *WHAT WAS THE WEATHER LIKE ON FRIDAY*. Since the key is 5 the first step involves writing the message in rows of 5 letters. This is:

$$
\begin{array}{ccccc}
W & H & A & T & W \\
A & S & T & H & E \\
W & E & A & T & H \\
E & R & L & I & K \\
E & O & N & F & R \\
I & D & A & Y
\end{array}
$$

Since the length of the message is not a multiple of 5, we must add one *Z* to get:

```
WHATW
ASTHE
WEATH
ERLIK
EONFR
IDAYZ
```

We now read down each column in turn to get the following cryptogram:

WAWEEIHSERODATALNATHTIFYWEHKRZ

To obtain the decryption key we merely divide the length of the message by the key. In this case we divide 30 by 5 to get 6. The deciphering algorithm is then identical to encryption. So, for this example, we write the cryptogram in rows of 6 to get:

```
WAWEEI
HSEROD
ATALNA
THTIFY
WEHKRZ
```

It is now easy to verify that reading down each column in turn gives the original message.

Transposition Ciphers of the type given here are easy to break. Since the key must be a divisor of the cryptogram length, an attacker has only to count the length of the cryptogram and try each divisor in turn.

Super-encryption

So far in this chapter we have provided a number of simple ciphers, most of which are easy to break. We now introduce a concept that can be used to combine two or more weak ciphers to obtain one that is considerably stronger than either. It is known as *super-encryption*. The basic idea is very simple. If, say, we want to super-encrypt using a Simple Substitution Cipher and a Transposition Cipher, then we first encrypt the message using the Simple Substitution Cipher, and then encrypt the resultant cryptogram using the Transposition Cipher. Once again a simple example should clarify the situation.

We encrypt the message *ROYAL HOLLOWAY* by super-encrypting a Caesar Cipher with key 2 with a Transposition Cipher with key 4. For the Caesar Cipher with key 2 we have:

Message: *R O Y A L   H O L L O W A Y*
Cryptogram: **T Q A C N   J Q N N Q Y C A**

For the Transposition Cipher with key 4 we have:

Message: *T Q A C N   J Q N N Q Y C A*
Cryptogram: **T N N A Q   J Q Z A Q Y Z C N C Z**

Super-enciphering is a very important technique and many modern strong encryption algorithms can be regarded as resulting from super-encryption using a number of comparatively weak algorithms.

Some conclusions

From the numerous examples discussed in the last sections it is clear that there are many factors that influence an attacker's chances of breaking a cipher system. We have also seen that, although the main prize for an attacker is knowledge of the decryption key, if the underlying language is highly structured then he may be able to determine a particular message without needing to discover the entire key. In fact our early examples indicate quite clearly that the structure of the underlying language is an extremely important factor when trying to assess an attacker's likelihood of success. For instance, it is clearly much easier to disguise random data than to successfully encrypt English text. It is also clearly true that, for English text, it is much easier to secure short messages than long ones. Indeed for a single short message, say three or four letters, there are many weak encryption algorithms that are probably sufficient.

Appendix

Introduction

In this appendix we discuss two basic mathematical ideas; binary representations of integers and modular arithmetic. Both play crucial roles in cryptography. Binary numbers are often taught in schools. Modular arithmetic is not so widely taught but, for special values like 7 and 12, it is a process which people do naturally.

Binary numbers

When writing an integer in decimal we use, essentially, a units' column, a tens' column, a hundreds' column, a thousands' column, and so on. Thus 3,049 is 3 thousands, 0 hundreds, 4 tens and 9 units. For decimal numbers we are working with base 10 and the columns represent the powers of 10, so $10^0 = 1$, $10^1 = 10$, $10^2 = 100$, $10^3 = 1,000$ and so on.

In binary numbers we work with base 2. The basic digits are 0 and 1 and we have a units' column, a twos' column, a fours' column (remember $4 = 2^2$), an eights' column ($8 = 2^3$) and so on. This now means that every binary string can be regarded as a number. For example, 101 in binary is 1 four, 0 twos, and 1 one, so the binary string 101 represents $4 + 0 + 1 = 5$ in decimal. Similarly 1011 in binary is 1 eight, 0 fours, 1 two and 1 one. Thus the binary string $1011 = 8 + 0 + 2 + 1 = 11$ in decimal. As one last example consider 1100011. This time we have 7 columns. The powers of 2 are 1, 2, 4, 8, 16, 32, 64 and so the binary string 1100011 represents $64 + 32 + 0 + 0 + 0 + 2 + 1 = 99$ in decimal.

Clearly any positive integer can be written in binary form and there are many ways of determining this form. We illustrate one method by doing a couple of examples. Suppose we wish to find the binary representation of 53. The powers of 2 are 1, 2, 4, 8, 16, 32, 64 . . . However, we can stop at 32 because all the other powers are bigger than 53. Now $53 = 32 + 21$. But $21 = 16 + 5$ and $5 = 4 + 1$. So $53 = 32 + 16 + 4 + 1$. All we have done is written 53 as the sum of powers of 2. We now note that $53 = (1 \times 32) + (1 \times 16) + (0 \times 8) + (1 \times 4) + (0 \times 2) + (1 \times 1)$ and so $53 = 110101$ in binary. For a second example consider 86. This time the highest power of 2 we need is 64. Repeating the earlier reasoning, we see that $86 = 64 + 16 + 4 + 2$. Thus 86 is 1010110 in binary.

The term *bit* is a contraction of binary digit. When we refer to an n-bit number, we mean that its binary form requires n bits. Thus, in the above examples, 53 is a 6-bit number and 86 is a 7-bit number. In general, $3.32d$ gives some idea of the number of bits needed to express d decimal digits in binary.

Modular arithmetic

Modular arithmetic is only concerned with integers, commonly known as whole numbers. If N is a positive integer, then arithmetic modulo N uses only the integers 0, 1, 2, 3, . . . , N-1, that is, integers from 0 to N-1.

There are a number of values of N for which arithmetic modulo N is common to most people, although they may not be familiar with the mathematical terminology. For instance when we use a 12-hour clock we use addition modulo 12. If it is now 2 o'clock then everyone 'knows' that the time in 3 hours is 5 o'clock, as it will also be in 15 hours. This is because $15 = 12 + 3$ and the time repeats every 12 hours. Other natural numbers are $N = 7$ (for the days of the week) and $N = 2$ (for odd and even).

If two numbers have the same remainder on division by N, we regard them as being equal modulo N. For example, if $N = 7$, then, since $9 = (1 \times 7) + 2$ and $23 = (3 \times 7) + 2$, we regard 9 and 23 as being equal modulo 7. If x and y are equal modulo N then we write x = y (mod N). Note then every integer must be equal modulo N to one of the values $0, 1, 2, \ldots, N\text{-}1$.

As an illustration of the use of modular arithmetic, suppose that the first day of a month is on a Tuesday. Then clearly the 2nd will be a Wednesday, the 3rd a Thursday, and so on. What about the 29th? One way of answering the question is to consult a calendar or to write out the days for the complete month. Another is to observe that the pattern of days repeats every 7 days. So, clearly, the 8th day is also a Tuesday. Now we note that $29 = 4 \times 7 + 1$ which, using the notation above, says $29 = 1$ (mod 7). Thus the 29th is exactly 4 weeks after the 1st and falls on a Tuesday. A similar argument shows that, since $19 = 2 \times 7 + 5$, the 19th will be a Saturday.

Once we have introduced the concept of working modulo N, then the arithmetic processes are straightforward. For example, if N = 11 then $5 \times 7 = 2$ (mod 11) since $5 \times 7 = 35 = (3 \times 11) + 2$. We can also write equations modulo N. So, for example, solve $3x = 5$ (mod 8) means find a value x such that the product of 3 with x is equal to 5 modulo 8. The solution to this particular equation is x = 7. It is beyond the scope of this appendix to provide methods to solve this type of equation. However, it is easy to verify that x = 7 satisfies $3x = 5$ (mod 8) because $3 \times 7 = 21 = 2 \times 8 + 5$. When talking about modular arithmetic it is important to remember that we are only 'allowed' to use integers, that is whole numbers. In particular if we are asked to solve an equation like $3x = 5$ (mod 8), then the answer must be an integer between 0 and 7.

One of the main reasons for including an introduction to modular arithmetic is that the two most popular public key algorithms use modular exponentiation as their basic mathematical process. Modular exponentiation means nothing more than computing x^a (mod N) for given integers x, a, N. Suppose x = 5, a = 4, and N = 7.

Then $5^4 = 5 \times 5 \times 5 \times 5 = 625 = (89 \times 7) + 2$, so $5^4 = 2 \pmod{7}$. There is no need for anyone using cryptography to be able to perform these calculations, but it almost certainly helps to understand the notation.

When discussing the Caesar Cipher we introduced additive ciphers, the notion of assigning numbers to the letters of the alphabet and then using arithmetic modulo 26. A close inspection of the Vigenère Square shows that the first row, used for key letter a, represents the additive cipher with shift 0, while the second row, determined by key letter b, represents an additive cipher with shift 1 and so on. In fact if we associate $A = 0$, $B = 1, \ldots, Z = 25$ then each key letter merely indicates the use of an additive cipher with shift equal to its associated value. This simple observation may prove helpful to anyone who tries to write a program to implement a Vigenère Cipher.

Our discussion of additive ciphers was followed by an introduction of the notion of a multiplicative cipher where we listed those numbers which can be used as keys. Although it may not be clear how we obtained the list, it is straightforward to check that the list is correct. For each of the numbers 1, 3, 5, 7, 9, 11, 15, 17, 19, 21, 23, 25, when we multiply the 26 letters of the alphabet by that number, we get 26 different answers. This means that the number can be used as the encryption key for a multiplicative cipher. The corresponding decryption keys are listed below and, once again, although it may not be apparent how they were calculated, it is straightforward to check that they are correct. The effect of multiplying by an encryption key and then multiplying by the corresponding decryption key should be to leave the letters unchanged, which is equivalent to multiplying by 1. For example, to check that, for an encryption key of 3, the decryption key is 9, we need only show that $3 \times 9 = 1 \pmod{26}$, which, since $3 \times 9 = 27 = 26 + 1$, is true.

Encryption key	1	3	5	7	9	11	15	17	19	21	23	25
Decryption key	1	9	21	15	3	19	7	23	11	5	17	25

Chapter 4
Unbreakable ciphers?

Introduction

The examples given in Chapter 3 were, of necessity, simple. Most of them are easily broken, although this was not necessarily the situation at the time they were designed. Cryptanalysis usually involves a considerable amount of trial and error, and recent advances in technology, especially computers, have facilitated this approach. The obvious example of a trial and error attack is the exhaustive key search discussed in Chapter 2. Trying all possible keys for a Vigenère Cipher with a reasonably long key word, of say six letters, would have been daunting when tried by hand in the sixteenth century. However, if we had a computer that could try 10,000 six-letter key words per second then it would take less than a day.

Before we move on from the historical examples of the last chapter to a discussion of modern techniques, it is worth discussing the concept of the unbreakable cipher. Being unbreakable is a claim that many designers have made for their algorithms, usually with disastrous consequences. We now give two famous historical examples of misplaced belief that a cipher was unbreakable, one from the 16th century and one from the Second World War.

Mary Queen of Scots used a variant of the Simple Substitution

Cipher in secret letters in the 16th century. This correspondence contained her plans both to escape from imprisonment and to assassinate Queen Elizabeth of England so that she could claim the English throne. The letters were intercepted, decrypted, and used as evidence in her trial. Mary and her conspirators openly discussed their plans in these encrypted letters, as they believed no one else would be able to read them. It was an error that cost Mary her life.

The German armed forces in the Second World War used a device called an Enigma machine to encrypt much of their important and unimportant military traffic. The mechanisms used by Enigma for encryption appear intricate and complicated, and a basic Enigma machine had over 10^{20} possible keys which is more than some modern algorithms. This led the users to believe that Enigma was unbreakable. As is now well known, the Allied Forces were at various times able to break Enigma, partially exploiting usage and key management mistakes. This effort was centred at Bletchley Park, which is now a museum. It is estimated that the work at Bletchley Park shortened the Second World War by two years.

In this chapter we discuss the concept of perfect secrecy, which, in some sense, is the best we can hope to achieve for the use of encryption. We then discuss the one-time pad, which is the only unbreakable algorithm.

Perfect secrecy

The general scenario that we have depicted so far is that of a sender trying to send a secret message to the intended recipient and using a cipher system to make the transmitted cryptogram unintelligible to a third party. Even if the third party fails to intercept that transmission, it is possible, although extremely unlikely in most situations, that they could actually guess the content of the message. Thus there is no way that the use of encryption can guarantee that the third party cannot obtain the content of the message. The best that the communicators can hope to achieve is that if the third party manages to intercept the transmission then that interception does not give them any information about the content of the message. In other words, the cipher system should be such that anyone who obtains the cryptogram is still forced to try to guess the message. However, there is no way that a cipher system can prevent attackers from trying to guess messages.

A system that achieves this objective is said to offer *perfect secrecy*, and we now give a small example to show that perfect secrecy is achievable.

Suppose that Mr X is about to make a decision that will have serious repercussions on the share value of a company. If he makes the decision 'buy' then the shares will increase in value, but the decision 'sell' will result in a collapse in their value. Suppose also that it is publicly known that he will soon be transmitting one of these two messages to his broker. Clearly anyone who received this decision before the broker would have the opportunity to use that information to either make a profit or avoid a disastrous loss, depending on the decision. Of course, at any time, anyone is free to guess what the message will be and to act accordingly. They have a 50 per cent chance of being right, and such an action would be nothing more than gambling.

Mr X wants to be able to send his decision over a public network as soon as he makes up his mind. Thus, in order to protect their interests, he and his broker decide to encrypt the message that will convey the decision. One option might be to use a Simple Substitution Cipher that, as we have already noted, is usually sufficient for protecting short messages. However, in this particular example, each message is uniquely identified by its length. Thus, on the assumption that the interceptor knows the system being used, knowledge of the length of the cryptogram would be sufficient to give the interceptor 100 per cent confidence that he knew the message, even though he could not determine the key used.

Another option might be to use the following system in which the two keys, k1 and k2, are equally likely. In order to describe the complete algorithm we use a standard notation. For key k1, the ciphertext for the plaintext message *BUY* is 0 while the ciphertext for *SELL* is 1. To express this simply we write $E_{k1}(BUY) = 0$ and $E_{k1}(SELL) = 1$. The expression $E_{k1}(BUY) = 0$ should be read as 'the result of encrypting *BUY* using key k1 is 0'. The complete cipher is:

$$\text{Key k1: } E_{k1}(BUY) = 0, E_{k1}(SELL) = 1$$
$$\text{Key k2: } E_{k2}(BUY) = 1, E_{k2}(SELL) = 0$$

An equivalent way of writing the same cipher is shown in the diagram.

	BUY	*SELL*
Key k1	0	1
Key k2	1	0

If this system is used and, say, a 0 is intercepted then all that the interceptor can deduce is that the message might be *SELL* if key k2 was used, or *BUY* if the key was k1. Thus the interceptor will be forced to guess which key was used and, since each key is equally likely to have been chosen, the chances of him guessing correctly are 50 per cent.

Note that, in essence, before the cryptogram was intercepted the attackers' only option was to try to guess the message. Once the cryptogram was received, they could also guess the key. Since the number of keys is the same as the number of messages, the chances of either guess being correct are equal. This is perfect secrecy. For this particular example the chances of an attacker guessing the message is 50 per cent, which is high. Thus, despite the fact that we have perfect secrecy, we have not provided any extra protection to increase the likelihood of the message remaining secret. Nevertheless the weakness is due to the fact that the number of messages is small. It is not the result of poor encryption.

There are a number of real-life situations where there are only a very limited number of potential messages and, in these situations, the risk of the message being guessed is greater than that of the encryption being compromised. An example that affects almost all of us is the use of PINs and credit or debit cards at an *Automated Telling Machine (ATM)*. In this scenario the user has a *Personal Identification Number (PIN)* that is used to establish that they own the card. If the PIN is verified at a financial institution's central computer, then encryption is used to protect it during transmission from the ATM to the host computer. If a user loses their card, then anyone who finds it can enter it into an ATM and enter a 'guessed' PIN value. Most PINs are four (decimal) digits, so there are at most 10,000 PIN values. The person who finds the card could then, in theory, keep guessing the PIN until they find the correct one, which would be easier then breaking the encryption. Furthermore there is no cryptographic solution to the problem. In recognition of this fact, most systems allow up to three false PIN entries before the ATM retains the card. This is one of many illustrations of when cryptography provides only partial solutions to problems, and *ad hoc* management decisions are necessary to increase the security.

It is, perhaps, worth noting here that, in our simple example of perfect secrecy, the agreement on which key was to be used could have been made as soon as the two parties knew they were likely to need to exchange secret data. This agreement could have been made in the privacy of one of their homes, and maintaining the secrecy of these keys could have been achieved by physical means, such as locking them in a safe until they were required. The relevance of this observation becomes apparent when reading about key management in Chapter 8.

Although our example of perfect secrecy had only two messages, it is possible to design similar schemes for any number of messages. However, perfect secrecy can only be obtained if the number of keys is at least as large as the number of messages.

The one-time pad

One fundamentally important consequence of our discussion on perfect secrecy is that it is attainable but, for systems where there are a large number of potential messages, only at the price of a potentially enormous key management overhead. The classic example of a perfectly secure cipher system is the *one-time pad*. If the message is a passage of English text containing n letters with all punctuation and spaces removed, then the key, which is only used once to protect a single message, is a randomly generated sequence of n letters from the alphabet. The encryption rule is precisely that used for the Vigenère Cipher with the key replacing the keyword. Thus, if we associate each letter A to Z with the numbers 0 to 25 in the usual way, for message $m_1, m_2, \ldots, m_n$ and key $k_1, k_2, \ldots, k_n$, the ith component of the cryptogram is given by:

$$c_i = (m_i + k_i) \bmod 26$$

Note that the fact that the key is the same length as the message guarantees that there is no need to start repeating the key during the encryption process.

There is another common version of this algorithm often called a *Vernam Cipher* where the alphabet used is binary, that is 0 and 1, and the cryptogram is obtained by adding the message and key modulo 2. For digital communications, the Vernam Cipher is almost certainly the version of the one-time pad that is used.

Since perfect secrecy is attainable, one might be tempted to ask why it is not used universally and why people use systems that may be broken. Before hinting at an answer to questions of this type, it is important to recall that the problems associated with the use of encryption for stored data tend to be different from those associated with its use to protect communications. It is also important to remember that we usually concentrate on communications, because these pose considerably more managerial problems.

When defining the one-time pad we restricted ourselves to listing the encryption algorithm and the encryption key. The decryption key is the same as the encryption key while the decryption algorithm involves subtracting the key characters for the cryptogram to obtain the original plaintext. The implementers of the communications system are now faced with a potentially difficult problem. How does the receiver obtain this random sequence? Since the sequence is randomly generated it is 'impossible' for the sender and receiver to generate the same key simultaneously. Thus one of them must generate the key and then send it (secretly) to the other party. If the secrecy of this key is to be guaranteed, then it needs protection during transmission. If the communicators have only one communications link available, then they need another one-time pad random sequence to protect the first one. Clearly this type of reasoning leads to the unachievable requirement of an infinite family of random sequences, each one being used to protect the previous one during its transmission from one party to the other. Thus one-time pads can only be used if the communicators have a second secure means of exchanging information. The reader will recall that Mr X and his broker had such a channel in our example of a perfectly secure system. It is also alleged that one-time pads are used for the very highest levels of secure links, such as the Moscow–Washington hotlines. In these situations a number of random sequences can be generated, stored, and then carried to the other locations by secure courier services. The sequences can then be stored in highly protected locations, produced when required and destroyed immediately after use. It is important to realize that this second secure channel is both slow and expensive. It could not be used for sending the messages since immediate responses and reactions may be required.

As we have already noted, key distribution in a secure network is a problem that is not restricted to the one-time pad. The need for a second secure channel is common. The difference is that whereas, for the one-time pad, the volume of traffic that it carries is comparable to the volume of message traffic, this second channel usually carries much less traffic. Even if there is a second, secure link, the one-time pad is unsuitable for systems that include many nodes, where each requires a secure link with every other. Here the problem is likely to be that of keeping track of which keys have been used and, possibly, handling the sheer volume of key material. Perfect secrecy depends upon each key being used only once. Thus the amount of key material required for a large, heavily used network is likely to make the overall key management totally impractical.

Not surprisingly, despite the fact that it offers the ultimate security level, there are very few practical communications networks systems that use the one-time pad. Of course, if someone is encrypting files to be stored for their own private use then they have no need to distribute any keys. In many storage scenarios, the only problems relate to key storage, and so, in some of these scenarios, the one-time pad may be as viable as any other cipher.

Chapter 5
Modern algorithms

Introduction

Throughout Chapter 3 we stressed that the examples given were not indicative of current practice and that modern encryption algorithms tend to operate on bits rather than the letter substitutions of our examples. In this chapter we discuss modern algorithms. Since they tend to be more complex than the examples of Chapter 3, we do not describe any specific examples in detail, but concentrate on the general techniques used to design them.

Bit-strings

As we stressed earlier, most modern ciphers do not involve letter substitutions. Instead they tend to use an encoding scheme to convert the message into a sequence of binary digits (bits), that is, zeros and ones. The most commonly used encoding scheme is probably ASCII (American Standard Code for Information Interchange). This bit sequence representing the plaintext is then encrypted to give the ciphertext as a bit sequence.

The encryption algorithm may act on a bit-string in a number of ways. One 'natural' division is between *stream ciphers*, where the sequence is encrypted bit-by-bit, and *block ciphers*, where the sequence is divided into blocks of a predetermined size. ASCII requires eight bits to represent one character, and so for a block cipher that has 64-bit blocks, the encryption algorithm acts on eight characters at once.

It is very important to realize that the same bit-string can be written in many different ways and, in particular, that the way that it is written may depend upon the size of the blocks into which it is divided.

Consider the following sequence of 12 bits: 1 0 0 1 1 1 0 1 0 1 1 0. If we split into blocks of size 3 we get: 100 111 010 110. However, any bit-string of length 3 represents an integer from 0 to 7 and thus our sequence becomes 4 7 2 6. For anyone who has skipped the Appendix to Chapter 3 and is unfamiliar with binary representations of integers:

$000 = 0$, $001 = 1$, $010 = 2$, $011 = 3$, $100 = 4$, $101 = 5$, $110 = 6$, $111 = 7$.

If we take the same sequence and split it into blocks of size 4 we get: 1001 1101 0110. This time, since bit-strings of length 4 represent the integers from 0 to 15, we get the sequence 9 13 6. In general a binary sequence of length n can be regarded as representing an integer from 0 to $2^n - 1$ and so, once we have agreed a block length s, an arbitrarily long binary sequence may be written as a sequence of integers in the range 0 to $2^s - 1$.

The precise detail of the mathematics is not important. However, it is important to observe that the same bit-string can be represented as a sequence of numbers in many different ways, depending upon the block size that is chosen. It is also important to remember that, if a block size is specified, for small numbers, it may be necessary to include some extra 0s at the beginning. For example, the binary representation of the integer 5 is 101. However, if we are using a block size of 6 then we represent 5 as 000101 and, for block size 8, 5 is 00000101.

Another common way of writing bit-strings is to use the *hexadecimal notation* (*HEX*). For HEX the string is divided into blocks of size 4 with the following representation:

$$0000 = 0 \quad 0001 = 1 \quad 0010 = 2 \quad 0011 = 3$$
$$0100 = 4 \quad 0101 = 5 \quad 0110 = 6 \quad 0111 = 7$$
$$1000 = 8 \quad 1011 = 9 \quad 1010 = A \quad 1011 = B$$
$$1100 = C \quad 1101 = D \quad 1101 = E \quad 1111 = F.$$

Thus the HEX representation of the sequence above is 9 D 6.

Since cipher algorithms operate on binary strings we need to be familiar with a commonly used method of combining two bits called *Exclusive OR* and often written as *XOR* or $\oplus$. It is identical to addition modulo 2 and is defined as follows: $0 \oplus 0 = 0$, $0 \oplus 1 = 1$, $1 \oplus 0 = 1$, and $1 \oplus 1 = 0$. This can be represented as a table.

	0	1
0	0	1
1	1	0

The table for the XOR or $\oplus$ operation

This simple operation gives us a way of combining two bit-strings of the same length. We XOR the pairs of bits in identical positions. So, for example, suppose that we wish to evaluate $10011 \oplus 11001$. The left-hand bit of 10011 is 1 and the left-hand bit of 11001 is also 1. Thus, since the left-hand bit of $10011 \oplus 11001$ is obtained by XORing the left-hand bits of the two individual strings, we note that the left-hand bit of $10011 \oplus 11001$ is $1 \oplus 1$ which is 0. Continuing in this manner we have $10011 \oplus 11001 = 1 \oplus 1 0 \oplus 1 0 \oplus 0 1 \oplus 0 1 \oplus 1 = 01010$. Another way of writing the same calculation is shown in the diagram.

Stream ciphers

Various authors use the term stream cipher slightly differently. Many talk about word-based, or character-based, stream ciphers. Here the message is enciphered word by word (or character by character), where the rule for the encryption of each word (character) is determined by its position in the message. The Vigenère Cipher, which was discussed in Chapter 3, and the one-time pad both fit this definition. Perhaps the best-known historical example is the celebrated Enigma cipher. However, the most common modern use of the term *stream cipher*, and the one that we adopt, is a cipher where the plaintext is enciphered bit by bit. Clearly all that can happen to any particular bit is that its value is changed to the alternative value or left unchanged. Since a bit can have one of only two values, changing a bit means replacing it by the other value. Furthermore, if a bit is changed twice, then it returns to its original value.

If an attacker knows that a stream cipher has been used, then their task is to try to identify the positions of those bits which have been changed, and to change them back to their original values. If there is any easily detectable pattern that identifies the changed bits then the attacker's task may be simple. Thus the positions of the changed bits must be unpredictable to the attacker but, as always, the genuine receiver needs to be able to identify them easily.

For a stream cipher we can think of the encipherment process as a sequence of the following two operations: change and leave unchanged. This sequence is determined by the encryption key and is often called the *keystream sequence*. For simplicity and brevity, we can agree to write 0 to mean 'leave unchanged' and 1 to mean 'change'. We are now in the position where the plaintext, ciphertext and keystream are all binary sequences.

In order to clarify our description, suppose that we have the plaintext 1100101 and the keystream is 1000110. Then, since a 1 in the keystream means change the bit of the plaintext that is in that position, we see that the 1 that is in the left-most position of the plaintext must be changed but that the next bit remains unchanged. Repeating this argument gives 0100011 as the ciphertext. We have already noted that changing a bit twice has the effect of returning it to its original value. This means that the decryption process is identical to the encryption process, so the keystream also determines decryption.

All that we have done in the above discussion is 'combined' two binary sequences to produce a third binary sequence by a rule which, in our particular case, can be stated as 'if there is a 1 in a position of the second sequence then change the bit in that position of the first sequence'. This is exactly the operation XOR or $\oplus$ which we defined in the last section. Thus if P_i, K_i and C_i are respectively the plaintext, keystream and ciphertext bits in position i, then the ciphertext bit C_i is given by $C_i = P_i \oplus K_i$. Note that decryption is defined by $P_i = C_i \oplus K_i$.

Stream ciphers are essentially practical adaptations of the Vernam Cipher with small keys. The problem with the one-time pad is that, since the keystreams are random, it is impossible to generate the same keystream simultaneously at both the sender and receiver. Thus they require a second secure channel for distributing the random keys and this channel carries as much traffic as the communications channel. Stream ciphers have the same requirement for a secure key channel but for considerably less traffic.

A stream cipher takes a short key to generate a long keystream. This is achieved by using a binary sequence generator. Note that in our discussion of the Vigenère Cipher in Chapter 3 we introduced the concept of using a generator to produce a long alphabetic keystream from a short alphabetic key. In that case, however, the generation process was very crude because it merely took a key word and kept on repeating it. Keystream generators for practical stream ciphers need to be more sophisticated than that. As an indication of why we note from above the keystream bit in position i, $K_i = P_i \oplus C_i$ can be determined as the XOR of the plaintext and ciphertext bits in position i. This highlights a potential weakness for stream ciphers because anyone who is able to launch a known plaintext attack can deduce parts of the keystream sequence from the corresponding plaintext and ciphertext bit pairs. Thus, users of stream ciphers must protect against attacks where the attacker deduces part of the keystream. In other words the keystream sequence must be unpredictable in the sense that knowledge of some of it should not enable an attacker to deduce the rest. For example a Vigenère Cipher with a short key of length 4 produces a keystream that repeats every four characters. However, it is straightforward to design keystream generators, which, for a suitably chosen four-bit key, repeat every fifteen bits. To do this we start with any key of length 4 except 0000. One generation process is then that each bit of the sequence is obtained by XORing the first and last bits of the previous four. If we start with 1111 the sequence continues 111101011001000 and then repeats forever. In fact, it is straightforward to take a key of length n and produce a keystream that does not start to repeat until $2^n - 1$ bits.

Designing good keystream sequence generators is quite difficult and some advanced mathematics is required. Furthermore, intense statistical testing is needed to ensure that, as far as is possible, the output from a generator is indistinguishable from a truly random sequence. Despite this, there are a number of applications for which stream ciphers are the most suitable type of cipher. One reason is that if a ciphertext bit is received incorrectly, then there is only one incorrect decrypted bit, as each plaintext bit is determined by only one ciphertext bit. This is not true for block ciphers where a single ciphertext bit received incorrectly results in the entire deciphered block being unreliable. This lack of 'error propagation' by the decryption algorithm is necessary if the ciphertext is being transmitted over a noisy channel and, as a result, stream ciphers are used for encrypting digitized speech such as in the GSM mobile phone network. Other advantages that stream ciphers tend to have over block ciphers include speed and ease of implementation.

Block ciphers (ECB mode)

For a *block cipher*, the bit-string is divided into blocks of a given size and the encryption algorithm acts on that block to produce a cryptogram block that, for most symmetric ciphers, has the same size.

Block ciphers have many applications. They can be used to provide confidentiality, data integrity, or user authentication, and can even be used to provide the keystream generator for stream ciphers. As with stream ciphers, it is very difficult to give a precise assessment of their security. Clearly, as we have already seen, the key size provides an upper bound of an algorithm's cryptographic strength. However, as we saw with the Simple Substitution Ciphers, having a large number of keys is no guarantee of strength. A symmetric algorithm is said to be *well designed* if an exhaustive key search is the simplest form of attack. Of course, an algorithm can be well designed but, if the number of keys is too small, also be easy to break.

Designing strong encryption algorithms is a specialized skill. Nevertheless there are a few obvious properties that a strong block cipher should possess and which are easy to explain. If an attacker has obtained a known plaintext and ciphertext pair for an unknown key, then that should not enable them to deduce easily the ciphertext corresponding to any other plaintext block. For example, an algorithm in which changing the plaintext block in a known way produces a predictable change in the ciphertext, would not have this property. This is just one reason for requiring a block cipher to satisfy the *diffusion property* which is that a small change in the plaintext, maybe for example in one or two positions, should produce an unpredictable change in the ciphertext.

We have already discussed the threats posed by exhaustive key searches. During such a search it is likely that the attacker tries a key that differs from the correct value in only a small number of positions. If there were any indication that the attacker had, for instance, tried a key which disagreed with the correct key in only one position, then the attacker could stop his search and merely change each position of that specific wrong key in turn. This would significantly reduce the time needed to find the key and is another undesirable property. Thus block ciphers should satisfy the *confusion property* which is, in essence, that if an attacker is conducting an exhaustive key search then there should be no indication that they are 'near' to the correct key.

When we discussed attacks on Simple Substitution Cipher, we gave an attack that gradually constructed the encryption key by first finding the substitute for the letter **E**, then the substitute for the letter **T** and so on. If an attacker is able to determine parts of the key independently of other parts, then they are said to be launching a divide-and-conquer attack. To prevent this we require *completeness*, which means that each bit of ciphertext should depend on every bit of the key.

Statistical testing forms a fundamental component of the assessment of block ciphers for these three listed properties and others. Thus statistical testing is vital for analysing all symmetric ciphers.

The simplest, and perhaps the most natural, way to apply a block cipher to a long message is to divide the binary sequence into blocks of appropriate size and then to encrypt each block individually and independently. When this occurs we say we are using the *Electronic Code Book* mode or *ECB* mode. When a key is chosen and ECB mode is used, then identical message blocks result in identical cryptogram blocks. This means that if an attacker ever obtains a corresponding plaintext and ciphertext block pair, they are able to locate that plaintext block everywhere in the message by finding the corresponding ciphertext bit. It is therefore worthwhile for him to build up a dictionary of known corresponding plaintext and ciphertext blocks. Furthermore, if there are popular message blocks then these result in equally popular cryptogram blocks. This might lead to the same type of frequency-based attack that we used on Simple Substitution Ciphers. This is one of the motivations for having comparatively large block sizes, such as sixty-four bits that, typically, correspond to eight characters. The use of ECB has yet another potential disadvantage, which we illustrate with an example.

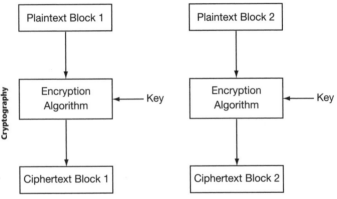

A block cipher in ECB mode

Suppose that an unknown block cipher with an unknown key is used to encrypt the message 'The price is four thousand pounds'. All that is known is that a message block consists of two letters, that punctuation, spaces, etc., are ignored, and the cryptogram is:

$$c_1, c_2, c_3, c_4, c_5, c_6, c_7, c_8, c_9, c_{10}, c_{11}, c_{12}, c_{13}, c_{14}$$

Suppose that an attacker knows the message. Then they are able to work out that c_1 represents Th, c_2 represents ep, etc. They can then manipulate the cryptogram so that only $c_1, c_2, c_3, c_4, c_5, c_6, c_7, c_{12}, c_{13}, c_{14}$ is received. The receiver applies the decryption algorithm with the correct key to the received cryptogram to obtain '*The price is four pounds*'. Since the decryption worked and the message makes sense, the receiver has no reason to suspect that the cryptogram has been manipulated and assumes that the price is correct.

Each of these potential dangers of using a block cipher in ECB mode can be removed by arranging for the encryption of each individual block to depend on all the message blocks that precede it in the message. If this is done, then identical message blocks almost certainly give distinct cryptogram blocks, and manipulation of the cryptogram is likely to result in meaningless messages after decryption has been applied. There are two standard ways of effecting this. They are known as *Cipher Feedback* (*CFB*) mode and *Cipher Block Chaining* (*CBC*) mode and are discussed later.

In order to illustrate how block ciphers are used in ECB mode we include a small example. The algorithm used is, of necessity, weak. In our example, the plaintext blocks, ciphertext blocks and keys are all of size 4 bits, and we use the HEX notation to describe them. For any given key K, the ciphertext block **C** corresponding to plaintext block M is obtained by XORing M with K and then rotating the bits of $M \oplus K$ one position to the left.

We encrypt the plaintext bit string 10100010001110101001, which becomes A23A9 when HEX notation is used with key K = B. The process is as follows:

Remember that we are using the HEX notation, so for the first block M = 1010 and K = 1011. Thus $M \oplus K$ = 0001. If we now perform the rotation we see that the ciphertext block is 0010, which is 2 in HEX.

Similarly for the second block if M = 2 and K = B. Thus M = 0010, K = 1011 and so $M \oplus K$ = 1001. If we now perform the rotation on 1001 we see that the ciphertext block is 3 in HEX.

Repeating this type of calculation we see that if the message is A23A9 and we use our cipher in ECB mode with K = B then the cryptogram is 23124.

The obvious observation is that the repeated block of the message results in a repeated block in the cryptogram.

Hash functions

So far we have concentrated on encryption algorithms that can be used to provide confidentiality. These algorithms have the fundamental basic property that they are reversible in the sense that, with knowledge of the appropriate key, it must be possible to reconstruct the plaintext message from the cryptogram. However, there are many instances where cryptography is used, but where there is no need to be able to deduce the original 'message' from its encrypted form. In fact there may be a definite requirement that it should not be possible. One example might be the protection of passwords on a computer system. Users are instructed to keep their passwords secret, and thus it is reasonable to assume that the system also tries to ensure this confidentiality. Thus, whenever the passwords appear in the system, notably on the database that is used for verification, they must be secured. However, the requirement here is usually to be able to verify that an entered password is correct and there may be no need to be able to deduce the password from the stored value.

There are also many instances in cryptography where large messages need to be condensed into short bit-strings (considerably shorter than the length of the original message). When this occurs, then it is inevitable that more than one message can give rise to the same shorter bit-string and this automatically implies that the process is irreversible. These functions are known as *hash functions* and, depending on the application, they may or may not involve the use of a cryptographic key.

The basic idea of a hash function is that the resultant hash value is a condensed representative image of the message. The hashed value has a number of names including *digital fingerprint*, *message digest*, or, not surprisingly, a *hash*. Hashing has a number of applications, including the provision of data integrity and as part of the digital signature process.

In general hash functions accept inputs of arbitrary length but produce outputs of a fixed length. If two inputs result in the same output then we say that a *collision* has occurred. As we have already noted, the existence of collisions is inevitable. Thus, if we want to identify a message uniquely by its digital fingerprint, the hash function must be carefully chosen to ensure that, even though collisions exist, it is virtually impossible to find them. This has a number of implications, one of which is that the number of possible fingerprint values must be large. To illustrate why we look at a ridiculously small example. If there were only eight possible fingerprint values then there would be a 12.5 per cent chance that two arbitrary messages have the same value. Furthermore, any collection of nine or more messages is guaranteed to contain at least one collision.

Public key systems

We have so far considered only symmetric algorithms where the sender and receiver share a secret key. This, of course, implies trust between the two parties. Prior to the late 1970s, these were the only algorithms available.

The basic idea of a public key cryptosystem is that each entity has a *public key* and a corresponding *private key*. These keys are chosen so that it is practically impossible to deduce the private key from the public key. Anyone wishing to use this system to send a secret message to someone else needs to obtain that person's public key and use it to encrypt the data. It is, of course, necessary that they have confidence that they are using the correct public key because, otherwise, it is the owner of the private key corresponding to the public key used, as opposed to the intended recipient, who can understand the message. Thus, although there is no need to distribute them secretly, all public keys need protection in the sense that their authenticity must be assured. It is also worth observing that when a public key system is used to provide confidentiality then, since the public encryption key is widely known and can be used by everyone, the ciphertext does not provide any authentication of the sender.

For a public key system both the algorithm and the encryption key are public. Thus an attacker is faced with the task of trying to deduce the message from the cryptogram, which was obtained by a method of which he has complete knowledge. Clearly the encryption process needs to be chosen very carefully to ensure that the attacker's task is difficult. However, it must not be forgotten that the genuine receiver needs to be able to decrypt easily. Thus the process must be chosen so that knowledge of the decryption key facilitates the determination of the message from the cryptogram.

This is a difficult and unintuitive concept. 'If everyone knows what I have done to determine the cryptogram then why can't they work out the message?' is a frequently asked question. The following non-mathematical example often helps.

Suppose you are in a closed room with no access to a telephone and are given a paper copy of the London telephone directory. If someone gives you a name and address and asks you for a telephone number, then this is an easy task. However, suppose you were given a randomly chosen telephone number and asked for the name and address of its owner. This, surely, would be a daunting task. The reason is not because you do not know what to do. In theory you could simply start at the first page and read all numbers until you find the correct one. The difficulty lies in the sheer volume of work involved. So if we regard 'name and address' as the message, the 'telephone number' as the cryptogram, and 'find the number of' as the encryption process, then for the London telephone directory, we have achieved the objective. It is important to note here that if the same process is used on smaller directories, then an attacker is able to reverse it. Furthermore it is not possible to be precise about how large a population is required before we feel justified in claiming that we achieve our objective. The London telephone directory has more than 750,000 names and we are happy to assert that in this context 750,000 is a large number. For an office block with only 100 extensions, reversing a directory is probably easy. What about a directory with 5,000 numbers?

There are, of course, organizations such as the emergency services that can identify the owners of arbitrary phone numbers. They have the directory listed in number order. Once again there is nothing to prevent anyone constructing their own version in number order. It is the magnitude of the task that ensures they do not succeed under the conditions we defined. However it is, of course, true that anyone with an electronic version of the directory would have a much simpler task.

Most practical public key algorithms are block ciphers that regard the message as a sequence of large integers and rely on the difficulty of solving a particular mathematical problem for their security. The most well-known was invented by Ron Rivest, Adi Shamir and Len Adleman in 1978 and is known as RSA. For RSA, the associated mathematical problem is factorization. There is a publicly known number N that is the product of two primes, whose values are secret. These primes are very important, because anyone who knows their values can use them to calculate the private key from the public key. Thus the number N, which determines the message block size, must be large enough that no attacker can deduce the primes, that is, no attacker can factorize N. Clearly, if N is small, anyone can determine the primes. A ridiculously small example is $N = 15$, when the primes are 3 and 5. However, it is believed that for large enough N finding the primes is infeasible. The difficulty of factoring large numbers is discussed in Chapter 7. For the moment we merely note that the number N determines both the block and key sizes.

This means that the sizes of the keys and blocks are likely to be much larger than those for symmetric ciphers. For symmetric ciphers, typical block sizes are 64 or 128 bits, whereas for RSA they are likely to be at least 640 bits and blocks of 1,024 or 2,048 are not uncommon. Another consequence is that the encryption and decryption processes involve many calculations using large numbers. This means that they are considerably slower than most symmetric algorithms. Thus they tend not to be used for encrypting large volumes of data but, instead, are more likely to be used for digital signatures or as key-encrypting-keys to distribute, or store, keys for symmetric algorithms.

The other widely used public key algorithm is known as El Gamal, which forms the basis of the U.S. *Digital Signature Standard* (*DSS*). For El Gamal, the sizes of keys are roughly the same as for RSA, but the security relies on the difficulty of a different mathematical problem known as the discrete logarithm problem. However, El Gamal possesses certain properties that do not make it well-suited for encryption.

The principles and standard techniques of public key cryptography had been developed in the early 1970s by James Ellis, Clifford Cocks and Malcolm Williamson at the U.K. government's Communications-Electronic Security Group (CESG). However, this work was classified for over two decades and was not made public until long after the initial public key cyptography papers, by which time asymmetric techniques had become highly developed.

Chapter 6
Practical security

Introduction

The term 'strong encryption' is widely used but, not surprisingly, can mean different things to different people. Typically it is often taken to mean 'unbreakable encryption', though this in itself is rather more subjective than might be expected.

It has been known for many years that the one-time pad is essentially the only unbreakable cipher. This was proved by Claude Shannon in two seminal papers in 1948 and 1949. These papers are the foundation of the modern theory of communications, including cryptography. It is impossible to overstate the significance of Shannon's contribution.

We have already seen that, for most practical systems, the use of the one-time pad is not feasible. Thus most practical systems use algorithms that are theoretically breakable. However, this need not imply that they are insecure. If, for example, all the theoretical attacks on the algorithm were too difficult to implement, then users might feel justified in regarding their algorithm as being effectively unbreakable. Even if this were not the case, in a particular application, the resources needed to break the algorithm might vastly exceed the value of the potential gain for any attacker. For this particular application, the algorithm could be regarded as 'secure enough'. Suppose, for example, that someone intends to use encryption to provide confidentiality for some data. They must then try to assess the value of the data being protected. This may not be straightforward. The value of the data may not be monetary, but may be purely personal. Obvious examples of data where it may be impossible to attach any quantitative values are medical records or other personal details. They must also make some sort of assessment about who is likely to want to gain access to their data and why. Other significant factors are likely to be the time for which the data must be kept secret, as well as the cost, availability, and ease of use of the algorithm.

When integrating cryptography into a security solution, there are two potentially conflicting approaches to the selection of an encryption algorithm:

- use the lowest level of security that offers adequate protection;
- use the highest level of security that implementation considerations allow.

Clearly, it is important for implementers to have a good idea of the level of security offered by an algorithm. This is discussed in the later sections of this chapter. The discussion concentrates mainly on exhaustive key searches for symmetric algorithms, and on attacks on the underlying mathematics for public key systems. Of course, as we have already stressed, the time for an exhaustive key search provides only an upper bound for an algorithm's strength. There may be other attacks which are easier. However, the design of algorithms is, we believe, sufficiently advanced for there to be numerous encryption algorithms that are well designed, in the sense that an exhaustive key search represents the easiest known form of attack. Furthermore implementation of these algorithms is likely to be very fast.

In the past, implementation considerations often forced users to adopt the policy of using the lowest level of security that they could.

Advanced technology then caught up with them, often with disastrous consequences.

Realistic security

Shannon showed that, essentially, the one-time pad is the only perfectly secure cipher system. Thus we know that, at least in theory, most practical systems can be broken. This does not, however, imply that most practical systems are useless. A (theoretically breakable) cipher system could be suitable for an application, if the users could be confident that it was very unlikely for an attack to succeed before the cover time for that application had elapsed.

The exhaustive key search is one basic form of attack that we have already discussed. The estimated time needed for an exhaustive key search being considerably longer than the cover time is one obvious 'hurdle' for a system to pass before it can be deemed to be suitable for a specific application. Of course, we have already seen that having a large number of keys is no guarantee of a secure system and, consequently, passing this requirement can only be regarded as the first of many tests before a system is deemed to be acceptable. Nevertheless failing this test is a clear indication that the algorithm cannot be used. Thus our first 'test' of a cipher system is to try to determine that the time needed for an exhaustive key search is sufficiently long, or equivalently that the number of keys is sufficiently large.

In order to do this, the designer needs to make a number of assumptions about the attacker's resources and capabilities. Their first task is to try to estimate how long an attacker is likely to require to try a single key. Clearly this time depends upon whether the attacker is using hardware or software. For a hardware attack, an attacker might use a purpose-built device. Any underestimation of this time is likely to lead to insecurity, whilst overestimating may make security a larger overhead than is needed.

A lucky attacker who attempts an exhaustive key search might find the key with the first guess. One result of having a large number of keys is to make the probability of this happening very small. At the other extreme, a very unlucky attacker might not find the key until the last guess. In practice, it is very unlikely that the attacker would need to perform a complete search before finding the key. The expected time for finding the key in a key search is likely to be much nearer to half the time required for a complete exhaustive search.

It is perhaps worth noting at this stage that if, for instance, an attacker has sufficient data, then they may be confident that only one key can transform all the known plaintext onto the correct ciphertext. However, in many situations the completion of an exhaustive search may not have identified the correct key uniquely. Instead it has reduced the number of potential candidates for the correct key and further searching with more data is required.

Once the number of keys has been fixed, the time needed for an exhaustive key search gives an upper bound for the security level. In many situations the designer's main aim is to try to ensure that the expected time for any other attack to succeed is longer than this bound. This is by no means an easy task. We have used words that indicate that time is the correct measure for assessing the likelihood of an attack succeeding. However the time needed for any computation depends upon a number of variables including, for instance, the processing power available and the technical/ mathematical ability of the attackers. The processing power available is related to the finances at the attacker's disposal and they, in turn, probably vary according to the expected profit resulting from a successful attack. Furthermore, in certain circumstances, other issues, such as the availability of computer memory to an attacker, are important. Bearing all this in mind, it is nevertheless true that such complexity measures are the normal way of determining whether or not any given system is secure enough for a particular application.

Practical exhaustive key searches

Although we do not want to include any complicated calculations, it is perhaps worth noting some facts so that we can get a 'feel' for the number of keys in certain situations. Clearly, anyone designing a system for a commercial application may want it to be secure for (at least) a few years and must therefore consider the impact of improving technology. This is often done by applying a very crude rule-of-thumb called *Moore's law* that states that the computing power available for a given cost doubles every 18 months.

In order to get some feel for sizes we note that there are 31,536,000 seconds in a year which is somewhere between 2^{24} and 2^{25}. If someone were able to try one key per second, then it would take them over a year to complete a search through 2^{25} keys. If, however, they had a computer capable of trying a million keys per second, the time required to search through 2^{25} keys would be significantly less than a minute. This is a dramatic change and, although very simplistic, is an indication of the impact that the advent of computers has had on the number of keys needed for secure systems. When discussing cryptographic algorithms some authors refer to the size of a key and others refer to the number of keys. It is worth recalling that there are 2^{s} bit patterns of length s which means that, if every possible bit pattern represents a key, saying that a system has s-bit keys is equivalent to saying that it has 2^{s} keys. It is also worth noting that, if every possible bit pattern represents a key, adding only a single extra bit to the key size has the effect of doubling the number of keys.

The most well-known symmetric block cipher is the *Data Encryption Standard (DES)*. It was published in 1976 and has been widely used by the financial sector. DES has 2^{56} keys and, ever since it was first released, there have been arguments about whether or not it can be regarded as strong. In 1998, an organization called the Electronic Frontier Foundation (EFF) designed and built a dedicated piece of hardware to conduct exhaustive DES key searches.

The total cost was about $250,000 and it was expected to find a key in about five days. Although the EFF do not make any claims that they optimized their design, this has now become accepted as providing a yardstick of the current state of the art. Roughly speaking, for $250,000 it is possible to build a machine that completes a search through 2^{56} keys in about a week. We can now extrapolate from this by increasing the cost or increasing the number of keys and, after building in some factor like Moore's law, obtain very crude estimates of how long is required to search through a specified number of keys, for a specified expenditure, at any time in the near future.

In addition to purpose-built hardware, there have been a number of other public exhaustive search efforts that have typically used computing power accumulated in an open internet-based key search. The most significant was probably one that was completed in January 1999. It used a combination of the EFF hardware and Internet collaboration involving more than 100,000 computers and took less than a day to find a 56-bit DES key.

We have concentrated on key searches on DES because it is such a high profile algorithm. When it was designed in the mid-1970s it was considered to be strong. Now, only 25 years later, DES keys are being found in less than a day. It is worth noting here that the success of recent DES searches has not surprised either current DES users or the designers of DES, who recommended (in 1976) that it should only be used for 15 years. Most current users of DES now implement what is called Triple DES. Here the key is either two or three DES keys (112 or 168 bits). Triple DES encryption with two DES keys k_1 and k_2, is shown in the figure, where E and D represent encrypt and decrypt respectively.

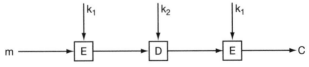

Two-key triple DES

In order to appreciate how dramatic the effect of adding eight extra bits to the key can be, we note that an internet-based search for a 64-bit key for an algorithm called RC5 was started early in 1998. After more than 1,250 days about 44 per cent of the possible keys had been tried and the correct one had not been found.

In 2001 the National Institute of Standards and Technology (NIST) published a new encryption algorithm that 'can be used to protect electronic data'. It is called the *Advanced Encryption Standard* (*AES*) and was chosen from a number of algorithms submitted in response to a request issued by NIST. The requirements were for a symmetric block cipher capable of using keys of 128, 192, and 256 bits to encrypt and decrypt data in blocks of 128 bits. The chosen algorithm is called Rijndael and was designed by two Belgians, Joan Daemen and Vincent Rijmen. Since AES has a minimum key size of 128 bits, it appears to be immune to exhaustive key searches using current technology.

We have already mentioned that Moore's law provides a rough estimate for improvements in existing technology improvements over the next few years. Moore's law is not concerned with radical new technologies that could have a dramatic effect. One such technology is quantum computing. Quantum computing performs calculations using quantum states that allow a form of parallel computation. Currently, only very small quantum computers have been built, and so they are essentially a theoretical concept. However, if quantum computers ever become a reality, then the situation will change dramatically. Very substantial sums of money are currently being spent worldwide on supporting research into the feasibility of quantum computing. If suitably sophisticated quantum computers could be built, then they would make exhaustive key searches significantly faster. As a rough guide they would double the size of the key length that could be searched in a given time. Thus, roughly speaking, a search through 2^{128} keys on a quantum computer would be as fast as searching through 2^{64} keys now.

Researchers are cautious about the likelihood of building quantum computers. Nevertheless there is some optimism in the field and the possibility should not be ignored.

Attacks on public key systems

Keys for asymmetric algorithms are longer than for symmetric ones. However, this does not mean that asymmetric algorithms are necessarily stronger. Exhaustive key searches are not the way to attack asymmetric algorithms. For an asymmetric algorithm, it is easier to attack the underlying mathematical problem. For example, for RSA, it is easier to try to factor the modulus N than to perform an exhaustive key search on all possible decryption keys.

In order to illustrate how recent mathematical advances have influenced the use of public key cryptography we concentrate on RSA and factorization. Similar observations apply to other public key systems, which depend on different mathematical problems.

The state of the art in factoring has advanced tremendously in the last 30 years. This is due to both theoretical and technological developments. In 1970, the 39-digit number $(2^{128} + 1)$ was factored into its two prime factors. At the time, this was considered a major achievement. When RSA was first published in 1978, the paper included a 129-digit number to be factored as a challenge, with a prize of $100. This was the first in a series of such challenges. This number was not factored until 1994, and the factorization involved the use of a worldwide network of computers.

In addition to Moore's law, the possibility of improved mathematical factoring techniques is another consideration when determining RSA key sizes. As an illustration we point to the dramatic impact caused by a mathematical innovation, called the *General Number Field Sieve* (*GNFS*), which was published in 1993. It meant that the resources needed to use previously known algorithms for factoring numbers of a given size could now be used to factor significantly larger numbers. For instance, resources that had been needed to factor a number of the order of 150 digits, might now factor a number nearer to 180 digits. This mathematical advance exceeded all performance increases predicted for technological advances for many years.

The 155-digit challenge number RSA-512 was factored using this technique in 1999. This factorization took less than eight months and, once again, involved a worldwide network of computers. An illustration of the mathematical complexity of the problem is that the final stage involved the solution of over six million simultaneous equations. This has been followed by a challenge published in *The Code Book*, which also required the factorization of a 512-bit modulus. These factorizations are significant as moduli of this size (155 digits or 512 bits) were routinely used in public key cryptography a few years ago.

Current recommendations for the moduli size for RSA typically range from 640 to 2,048 bits, depending on the security required. A 2,048-bit number has 617 decimal digits. To demonstrate how enormous this number is, we give the RSA challenge number of this size. Fame and a prize of $200,000 await the first team to successfully factor it.

```
25195908475657893494027183240048398571429282126204
03202777713783604366202070759555626401852588078440
69182906412495150821892985591491761845028084891200
72844992687392807287776735971418347270261896375014
97182469116507761337985909570009733045974880842840
17974291006424586918171951187461215151726546322822
16869987549182422433637259085141865462043576798423
38718477444792073993423658482382428119816381501067
48104516603773060562016196762561338441436038339044
14952634432190114657544454178424020924616515723350
77870774981712577246796292638635637328991215483143
81678998850404453640235273819513786365643912120103
9712288221 20720357
```

When discussing exhaustive key searches, we mentioned the potential impact of quantum computers. Although they would cause dramatic increases in the sizes of symmetric keys, there is little doubt that the cryptographic community would adapt and that symmetric algorithms would continue to be used securely. The same may not be true for public key systems. For these systems quantum computing would be a more serious threat. For instance factorization would become significantly easier. Fortunately even the most optimistic quantum computing enthusiasts are not predicting large quantum computers for at least 20 years.

Chapter 7
Uses of cryptography

Introduction

So far we have assumed that our cryptographic algorithms are being used to provide confidentiality. However, there are many other applications. Whenever we use cryptography it is important that we check that it is helping us achieve our desired objectives. We illustrate a potential misuse of cryptography with an example.

In 1983 MGM produced a film called *War Games*. It became a cult film that highlighted the dangers of hacking. One synopsis describes the film by saying 'The fate of mankind rests in the hands of a teenager who accidentally taps into the Defence Department's tactical computer'. Its opening scene showed the teenager hacking into his university's computer system and changing his girlfriend's grades. At that time, many universities were storing examination results on databases that could be accessed remotely. Not surprisingly, they were concerned that their results might be vulnerable to the type of unauthorized manipulation depicted in the film and wanted to introduce appropriate protection.

One proposal was to encrypt each student's grades. However, this did not really achieve the objective and it is both important and interesting to understand why. It is easy to see what encrypting the grades achieves. The result is that anyone who hacks into the database does not see the grade of any individual student. Instead they see meaningless data attached to each name. Unfortunately, this does not necessarily prevent constructive alteration of grades by hackers. If a hacker has failed, but happens to know that a specific student has good grades, then they merely change the meaningless data by their name so that it is identical to that associated with the other student. Of course, if they do not know the precise grades of the other student, then they do not know their new grades. Nevertheless, they know that they now have a pass grade. This is just one of many instances where the use of encryption fails to meet the user's objectives. It is not the answer to all problems. Note also that, in this particular example, the algorithm has not been broken. In fact it has not even been attacked. All that has happened is that the user failed to analyse the problem correctly.

Name	Encrypted grades
Good	13AE57B8
Bad	2AB4017E

→

Name	Encrypted grades
Good	13AE57B8
Bad	13AE57B8

OR EVEN

Name	Encrypted grades
Good	2AB4017E
Bad	13AE57B8

Suppose now that if, instead of encrypting only the grades, universities encrypted the complete database. Would this have achieved the objective of preventing a hacker changing grades? In this case, encrypting the complete database means that the complete file would be unintelligible to a hacker. However, even in this case, it may not be sufficient to protect against a hacker changing grades. Suppose, for instance, that each line of the file represented a student's name and grades. If the order in which the students appear is the alphabetical order of the complete class, then the attack discussed in the last paragraph would still be possible.

Before we concentrate on how cryptography might be used to protect stored information from being manipulated, we pause to consider whether or not it really matters if someone could alter the marks stored on any one particular database. It is, of course, absolutely crucial that students are credited with their correct grades. If the database being altered is not the only record available, then it might not be possible for the student to derive any benefit from the grades being changed in that particular record. The crucial requirement is probably that there should be some mechanism to warn all authorized users that the grades have been altered. Thus it may be that prevention of alteration is not vital, provided that any changes can be detected. This might mean that authorized users are warned not to rely on that database and to consult the master record. In many situations it is the detection of unauthorized changes that is required, rather than their prevention.

Cryptography is commonly used to ensure the detection of unauthorized alterations to documents. Indeed, at least for the commercial sector, the provision of confidentiality is no longer its major application. In addition to its traditional use for privacy, cryptography is now used to provide:

- *data integrity*: assurance that information has not been altered by unauthorized or unknown means;
- *entity authentication*: corroborating the identity of an entity;
- *data origin authentication*: corroborating the source of the information;
- *non-repudiation*: preventing the denial (this is usually by the originator) of the content of the information and/or the identity of the originator.

There are, of course, a number of standard (non-cryptographic) ways of protecting data from accidental corruption, for example, the use of a parity check or more sophisticated error-correcting codes. If, however, protection against deliberate alteration is required, then, since they depend only on public information, these techniques may not be sufficient. Anyone deliberately altering the information would encode the altered message appropriately so that the alteration was undetected. Thus, for protection against deliberate alteration, some value known only to the sender and (possibly) the receiver, such as a cryptographic key, must be used.

Using symmetric algorithms for confidentiality

We have already identified some potential security risks if a block cipher is used to encrypt data in ECB mode. One is the possibility that someone with knowledge of corresponding plaintext and ciphertext blocks could manipulate the ciphertext blocks to construct a cryptogram which would decrypt into a meaningful message. The changes would not be detected by the receiver. We have already seen a small example of this. However, the emphasis here should be on the word *meaningful*. If a block cipher is being used in ECB mode then, clearly, the decryption algorithm can be given the ciphertext blocks in any order and it would be able to decrypt each one individually to produce a potential message. However, it is unlikely that the resultant decrypted data would form a coherent, intelligible message. Although the possibility of such attacks should not be ignored, the chance of them succeeding is small.

The more serious deficiencies of using ECB mode are that an attacker might be able to build dictionaries of known corresponding plaintext and ciphertext blocks for a given key and that ECB mode is vulnerable to attacks which rely on the statistics of the underlying language of the plaintext. The 'classic' example of such an attack is the one shown for Simple Substitution Ciphers in Chapter 3.

Each of these vulnerabilities exists because the blocks are enciphered independently of each other. Thus, for a given key, identical plaintext blocks result in identical ciphertext blocks. One way of trying to overcome this is make each ciphertext block depend not only on the corresponding plaintext block, but also on its position in the complete text. This was the approach adopted in the Vigenère Cipher. Such techniques certainly have a 'flattening' effect on the statistics of the language. However, a more common and more effective technique is to ensure that the ciphertext block corresponding to any given plaintext block depends on the contents of all previous plaintext blocks in the message. The two most common ways of achieving this are the use of Cipher Block Chaining (CBC) and Cipher Feedback (CFB) modes. We describe Cipher Block Chaining (CBC) mode.

Suppose that we have a message consisting of n message blocks, $M_1, M_2, \ldots, M_n$ that we wish to encrypt with a block cipher using a key K. For CBC mode the resultant cryptogram has n blocks, $C_1, C_2, \ldots, C_n$, but now each of these cryptogram blocks depends on all previous message blocks. The way in which this is achieved is that each cryptogram block, other than C_1, is obtained by encrypting the XOR of the corresponding message block with the previous cryptogram block. Thus, for instance, C_2 is the result of encrypting $M_2 \oplus C_1$. So, if we write EK to represent the encrypting process with key K, we have $C_2 = EK(M_2 \oplus C_1)$. Clearly the first message block has to be treated differently. One option might be to let C_1 be $EK(M_1)$. Another common option is to use an initialization value (IV) and let C_1 be the result of encrypting $M_1 \oplus IV$. (Note that if IV is all 0s, then these two options are identical.) Since C_1 depends on M_1, and C_2 depends on M_2 and C_1, it is clear that C_2 depends on both M_1 and M_2. Similarly, since $C_3 = EK(M_3 \oplus C_2)$, C_3 depends on M_1, M_2, and M_3. In general, each ciphertext block depends on the corresponding plaintext block and all previous plaintext blocks. This has the effect of linking all the ciphertext blocks together in the correct specific order. It not only destroys the underlying statistics of the message, but also virtually eliminates the possibility of manipulation of the ciphertext.

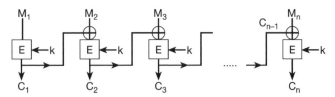

Cipher Block Chaining

We now illustrate how CBC mode works with the small example of a block cipher that we used in Chapter 5 and compare the cryptograms by using the same algorithm and key in EBC mode and CBC modes. Unfortunately the example looks more complicated than it really is. Readers are encouraged to persevere. However, you will not be disadvantaged if you jump to the beginning of the next section.

For this example, the plaintext, written in HEX, is A23A9 and the key K = B. The encryption algorithm XORs the plaintext block with the key and the ciphertext block is obtained by rotating the bits of $M \oplus K$ one position to the left. For CBC mode, we use IV of all 0s, so that C_1 is the same as for ECB mode. Thus C_1 is obtained by rotating $M_1 \oplus K = A \oplus B = 1010 \oplus 1011 = 0001$ to get 0010. So $C_1 = 2$.

To compute C_2 the process is as follows:

$$M_2 \oplus C_1 = 2 \oplus 2 = 0010 \oplus 0010 = 0000$$
$$0000 \oplus K = 0 \oplus D = 0000 \oplus 1011 = 1011$$

Performing the rotation gives $C_2 = 0111 = 7$. To compute C_3 we have:

$$M_3 \oplus C_2 = 3 \oplus 7 = 0011 \oplus 0111 = 0100$$
$$0100 \oplus K = 0100 \oplus 1011 = 1111$$

Performing the rotation gives $C_3 = 1111 = F$. To compute C_4 we have:

$$M_4 \oplus C_3 = A \oplus F = 1010 \oplus 1111 = 0101$$
$$0101 \oplus K = 0101 \oplus 1011 = 1110$$

Performing the rotation gives $C_4 = 1101 = B$. We leave the reader to compute C_5 (answer given below).

Thus, from the same message we obtain two cryptograms, depending on the mode of encryption.

Message: A 2 3 A 9
Cryptogram using ECB mode: 2 3 1 2 4
Cryptogram using CBC mode: 2 7 F B F

Even from our small example, it is clear that there is no obvious correlation between the positions of identical message blocks and the positions of identical ciphertext blocks.

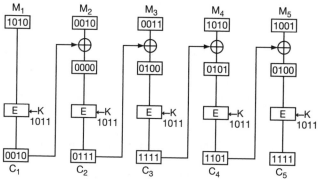

A diagram to illustrate the CBC example

When block ciphers are used in CFB mode, the process is different. However, the resultant effect is similar in the sense that each ciphertext block depends on the corresponding plaintext block and every preceding plaintext block in the order in which they appear in the message. For details of CFB mode see Menezes, van Oorschot, and Vanstone, *Handbook of Applied Cryptography*.

Authentication

The word *authentication* has two distinct meanings in the context of information security. One meaning relates to data origin authentication, which is concerned with verifying the origin of received data, while the other meaning relates to (peer) entity authentication, where one entity verifies the identity of another.

Data origin authentication is, typically, accompanied by a process of confirming the integrity of the data. Entity authentication takes many forms, but when based on cryptography, tends to rely on an exchange of messages between the pair of entities. This exchange is called an *authentication protocol*. Throughout this book we have frequently referred to users and regarded them as people. However in this context an entity is likely to be either a computer or a human user.

User authentication is, of course, fundamental to the concept of access control, and there are a number of ways for users to authenticate themselves, either to each other or to computer networks. The basic techniques tend to depend on at least one of the three following properties:

- *something known:* this might, for example, be a password or a PIN that the user keeps secret;
- *something owned:* examples include plastic cards or hand-held personalized calculators;
- *some characteristic of the user:* these include biometrics, such as fingerprints and retina scans, hand-written signatures or voice recognition.

Probably the most common methods involve combinations of something known or something owned. There is, of course, always the danger that anything known may be discovered by an adversary and that they might steal, or copy, anything that is owned. This lends support to the claim that the only methods that can genuinely authenticate users should depend on one of their characteristics, such as a *biometric*. However, for a number of practical reasons, biometrics are not yet widely implemented.

Using symmetric algorithms for authentication and data integrity

Authentication and data integrity can both be achieved using symmetric cryptography. We consider first authentication and then data integrity. There are two types of authentication. In *one-way authentication*, one user is authenticated to another user; whereas in *two-way authentication*, both users are authenticated to each other. Chapter 9 discusses the use of a magnetic stripe card at an ATM (cash machine), which is an example of one-way authentication. The card is cryptographically authenticated to the ATM by using a PIN. However, the cardholder has to use non-cryptographic means such as the location and design of the ATM to be convinced that the ATM is genuine. Logging onto a computer is usually another example of one-way authentication. Either type of authentication involves the use of an agreed algorithm and secret information or key. The correct use of this key in the algorithm provides the authentication. Clearly, this process relies on the key not being compromised. Furthermore, advanced authentication techniques often require the use of an agreed protocol involving the exchange of challenges and responses (which are encrypted versions of the challenge).

It must be noted that use of an authentication protocol only establishes the identities of the parties at the instant that the protocol took place. If either confidentiality or data integrity is required for information during the life of the connection that has just been authenticated, then other cryptographic mechanisms are needed to provide that protection. The keys needed for these cryptographic processes may be exchanged as part of the authentication protocol. However, if protection is needed against the replay of (part of) the authentication protocol by an impostor, then extra information, such as sequence numbers or time stamps, also have to be included.

Data integrity for a message can be assured using an authentication algorithm and a secret key. The authentication algorithm accepts the message and the agreed key as input and then calculates an authentication value which is the output. This authentication value is nothing more than a (short) bit-string whose value depends on the authentication algorithm, the message and the agreed key. In the terminology of Chapter 5, the authentication algorithm is a keyed hash function.

When user A wishes to send a message to user B, he appends the authentication value to the message. B receives the message and its authentication value. B then calculates the output of the authentication algorithm with the message received from A and the agreed secret key as input. If this output agrees with the authentication value sent by A, then B can be confident that the message came from A and has not been altered. (Thus the authentication function provides both data integrity and authenticates A.) The observant reader will have noticed that the use of this type of authenticator does not prevent replays. In order to protect against this type of attack, as we have already noted, users need to append identifiers, such as sequence numbers, to the messages.

One important aspect of this authentication process is that the sender and receiver perform exactly the same calculations. Thus, if there were ever a dispute between A and B as to what was sent, there would be no cryptographic way of settling it. This is not really a fault of the system, but merely a consequence of using symmetric cryptography. Here A and B must trust each other. They share a secret key and are relying on the secrecy of that key to protect them against alteration attacks from any third party. They are not seeking protection from each other because they have this mutual trust. In general this is true of most users of symmetric cryptography. It is

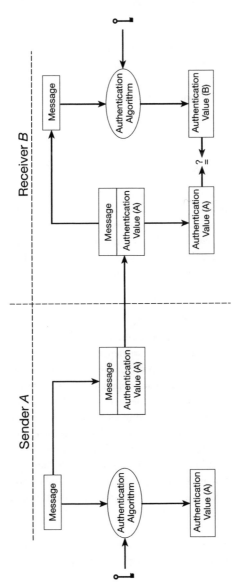

Authentication with symmetric authentication

used by mutually trusting parties to protect their information from the rest of the world.

The most widely used authenticator, particularly by the financial sector, is called a *Message Authentication Code* (*MAC*). If the message is $M_1, M_2, \ldots, M_n$, where each M_i consists of 64 bits, then DES in CBC mode is used. However, the only ciphertext block that is required is C_n. The MAC then consists of 32 bits of C_n.

Digital signatures

For the reasons explained in Chapter 5, the use of asymmetric algorithms tends to be restricted to the protection of symmetric keys and to the provision of digital signatures. If there is a requirement for settling disputes between sender and receiver as to the contents of a message or of its origin, then the use of symmetric cryptography does not provide the answer. Digital signatures are required.

The *digital signature* for a message from a particular sender is a cryptographic value that depends on the message and the sender. In contrast, a hand-written signature depends only on the sender and is the same for all messages. A digital signature provides data integrity and proof of origin (non-repudiation). It can be kept by the receiver to settle disputes if the sender were to deny the content of the message or even to deny having sent it. It is the provision of a means of settling disputes between sender and receiver that distinguishes the digital signature mechanism from the MACing process described in the last section. Clearly such disputes can only be settled if there is asymmetry between sender and receiver. This observation suggests asymmetric cryptosystems as the natural tools for providing digital signatures.

For a digital signature scheme based on a public key system such as RSA or El Gamal, the basic principle is very simple. Each user has a private key that only they can use, and its use is accepted as identifying them. However, there is a corresponding public key. Anyone who knows this public key can check that the corresponding private key has been used, but cannot determine the private key.

Accepting that the private key must have been used gives the receiver assurance of both the origin and content of the message. However, the sender is reassured that impersonation is impossible as the private or *signature* key cannot be deduced from the public or *verification* key or the digital signature.

Asymmetric cryptographic processing requires much computational processing. Thus a condensed version or hash of the message is produced by applying a hash function to the message. The signature is produced from the hash (which represents the message) by using the asymmetric algorithm with the private key. Thus only the owner of the private key can generate the signature. The signature can be verified by anyone who knows the corresponding public key. To do this a value is produced from the signature using the asymmetric algorithm with the public key. This value should be the hash of the message, which anyone can calculate. If this value and the hash agree, the signature is accepted as genuine. If they disagree, the signature is not genuine.

The two most widely used asymmetric algorithms are RSA and El Gamal. For RSA, the encryption and decryption are identical, so the signature and verification processes are also identical. An alternative to RSA is the Digital Signature Standard (DSS), which is based on El Gamal. For DSA, signature and verification processes are different. Furthermore, DSA requires a random number generator (extra processing), whereas RSA does not. However, DSA always produces a fixed length 320-bit signature. By contrast, for RSA the signature block and the modulus have the same size, which increases as the security level increases.

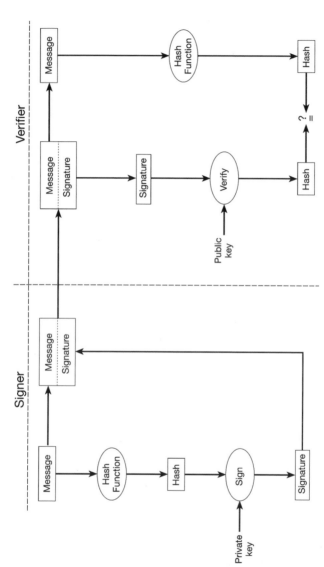

Digital signatures

Suppose digital signatures are being used as a means of identification. If user A wishes to impersonate user B, then there are two different forms of attack:

1. A attempts to obtain the use of B's private key.
2. A tries to substitute their public key for B's public key.

Attacks of the first type involve either trying to break the algorithm or gaining access to the physical devices that store the private key. Attacks on the algorithm have been discussed in Chapter 6; while the need for physical security is an important aspect of key management, which is the focus of Chapter 8. Both types of attack are similar to those that are launched at symmetric systems. However, attacks of the second type are unique to public key systems and most of the current 'defences' involve the use of digital certificates issued by Certification Authorities.

Certification authorities

We have already discussed 'traditional' attacks on cryptographic systems, such as breaking the algorithm to determine the private key, or obtaining access to the key by physical means, such as possession of a device enabling its use or probing to read the secret value. However, public key systems also need an infrastructure to prevent impersonation attacks. Suppose user A were able to establish their public key as belonging to user B. Other users would then use A's public key to encrypt symmetric keys for B. However A, as opposed to B, would obtain the secret information protected by these symmetric keys. Furthermore, A would be able to sign messages using their private key, and these signatures would be accepted as B's. The use of Certification Authorities and the establishment of Public Key Infrastructures (PKI) aim to prevent these impersonation attacks.

The main role of a *Certification Authority* (*CA*) is to provide digitally signed '*certificates*' that bind an entity's identity to the value of its public key. In order that the CA's certificates can be checked, the CA's own public key must be widely known and accepted. Thus, in this context, a certificate is a signed message that contains the entity's identity, the value of its public key, and probably some extra information such as an expiry date. These certificates can be thought of as a 'letter of introduction' from a well-respected source (the CA).

Suppose that CERTA is a certificate issued by the CA containing A's identity and A's public key, so CERTA binds A's identity to their public key value. Anyone with an authentic copy of the CA's public key can verify that the signature on CERTA is correct and thereby gain assurance that they know A's public key. Thus the problem of guaranteeing the authenticity of A's public key has been replaced by the need to be able to guarantee the authenticity of the CA's public key, coupled with trust that the verification of A's identity was performed correctly. Note that anyone who can impersonate A during the certification process can obtain a certificate binding their public key to A's identity. This enables them to impersonate A during the complete lifetime of that certificate. This is an example of the potentially worrying problem of identity theft that is likely to increase in the future.

It is important to note that anyone may be able to produce a given user's certificate so that ownership of user A's digital certificate does not identify A. The certificate merely binds A's identity to a public key value. Proof of identity can then be established by the use of a challenge–response protocol which proves the use of A's private key. This might involve A being issued with a challenge to sign. A returns it with their signature and the verifier confirms the validity of the signature by using the public key value in A's certificate. It is the use of the private key corresponding to the public key given in A's certificate that establishes A's identity.

Suppose now that two users, A and B, have certificates issued by different CAs. If A needs assurance about the authenticity of B's public key, then *A* needs an authentic copy of the public key of *B*'s CA. This might be achieved by *cross-certification*, where each of the two CAs issues a certificate for the other, or the introduction of a *certification hierarchy*, where a root CA sits above both those two CAs and issues certificates to each of them.

The diagrams illustrate the two processes. In each case *X* and *Y* are CAs while $X \rightarrow A$ implies that *X* issues a certificate for *A*. In (*b*) *Z* is a root CA. If, for example, *B* needs confidence in *E*'s public key, then for (*a*) *B* needs to verify the certificate for *Y* issued by *X* and the certificate of *E* issued by *Y*. For (*b*) *B* needs to verify the certificate of *Y* issued by *Z* and the certificate of *E* issued by *Y*. Thus, in each case, *B* needs to check a chain of two certificates. For more complex systems involving a combination of many cross-certifications and hierarchies with more than one level, this chain may be considerably longer.

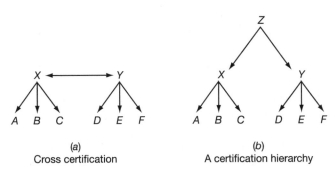

(a)
Cross certification

(b)
A certification hierarchy

Many regard digital signatures as being central to e-commerce, and many countries are introducing legislation so that they have the same legal status as hand-written signatures. For a recent comprehensive survey of digital signature mechanisms and a discussion of the issues associated with PKIs, we refer the reader to Piper, Blake-Wilson and Mitchell, *Digital Signatures*. However, some of the issues are very important and need to be identified here. One of the main problems associated with the use of certificates is the problem of *revocation*. For example, a company might issue a certificate to an employee who later leaves. A second example is that of a keyholder who knows that their private key has been compromised. In either case, there is a requirement for the CA to be able to revoke the certificate. Since these certificates are likely to have been widely distributed, it is unlikely to be practical to notify everyone directly. One common solution is for the CA to publish a *Certificate Revocation List* (*CRL*). However, this is a significant management overhead and has many associated problems.

A second obvious problem relates to liability. Many users are going to rely on such certificates. Suppose that a certificate is erroneous, in the sense that the public key value listed does not belong to the listed owner. In this case, it may not be clear who is liable: the owner, the user, or the CA.

Public Key Infrastructure

The motivation for using a PKI is to facilitate the use of public key cryptography. In Adams and Lloyd, *Understanding Public-Key Infrastructure*, which was, so far as we are aware, the first book devoted to the topic, *PKI* is defined as 'a pervasive security infrastructure whose services are implemented and delivered using public key concepts and techniques'.

We have already stressed the importance of the identity verification process, the need to be able to revoke certification, and the concept of cross-certification. Clearly cross-certification is likely to be extremely difficult unless the CAs involved are using compatible technologies. Even if they are, there are likely to be a number of issues associated with the general problem of how users can decide which CA's certification they can trust. Thus CAs need to publish policy and practice statements which contain, amongst other information, clear statements about their security procedures.

So far we have identified three key players in a PKI system. They are the certificate owner, who applies for the certificate, the CA, which issues the certificate that binds the owner's identity to the owner's public key value, and the relying party, who uses (and therefore relies) on the certificate. In some systems the identification verification is performed by a separate authority, called the *Registration Authority (RA)*.

As we have already seen, in a large PKI with numerous CAs the process whereby a user establishes the authenticity of another user's public key may involve verifying the signatures on a long chain of certificates. This can be an expensive, time-consuming exercise which users do not want to perform. In order to 'spare' users this overhead, the concept of a *Validation Authority (VA)* has emerged. The basic idea is that end users then merely ask the VA if a given certificate is still valid and receive a yes or no answer. The work is transferred from the user to the VA.

PKIs and digital signatures are arguably the two areas of cryptography of most current relevance to e-commerce. However, would-be implementers appear to be experiencing a number of technical difficulties; for example, problems associated with scalability. Furthermore, despite the fact that there are claims that PKI technology is crucial for secure email, secure web-server access, secure Virtual Private Networks, and other secure communications applications, the commercial inducement for an organization to set up as a CA has proved to be less attractive than anticipated.

When a PKI is established, the following processes need to take place, though not necessarily in the order listed:

- The key pairs for CAs must be generated.
- The key pairs for users must be generated.
- Users must request certificates.
- Users' identities must be verified.
- Users' key pairs must be verified.
- Certificates must be produced.

- Certificates must be checked.
- Certificates must be removed/updated (when necessary).
- Certificates must be revoked (when necessary).

Fundamental questions relating to these processes are *Where?* and *By whom?* Some CAs produce certificates which have different 'levels' attached to them where, loosely, the level is an indication of the extent to which the certificates may be trusted. So, for instance, users would be advised not to rely on low-level certificates for high value transactions. In such systems the level of a certificate is likely to reflect how the identification process was conducted. If, for example, identification was established by the use of an email address, then the resultant certificate would have a low level, while high level certificates might be issued only after a manual process including the production of a passport. For a well-written survey of the problems associated with PKIs and potential solutions we refer to Adams and Lloyd, *Understanding Public-Key Infrastructure*, or Clapperton, *E-Commerce Handbook*.

The need for trust

Certification Authorities provide one example of the concept of a *Trusted Third Party* (*TTP*). In this instance two parties trust a third party, the CA, and they then use this trust to establish a secure communication between themselves. TTPs appear almost everywhere that cryptography is used, and their use frequently causes concern. In general TTPs need to the trusted both for their integrity and also for their technical competence. It is often difficult to decide precisely how influential they should be and how much users' security should depend on them.

Consider, for example, the generation of public and private key pairs. As we have already noted, this is a mathematical process that requires dedicated software. It is not something that the ordinary citizen can do for themselves so the keys or the key-generation software are provided externally. In either case, there is an undisputed need for trust. Frequently the keys are generated externally. The obvious question is whether the keys should be generated by the CA or by another TTP. Our aim is not to provide an answer as this clearly depends on both the application and environment, but merely to draw attention to some of the issues. The fear is that, if an organization generates the private and public key pair of another entity, they might keep a copy of the private key or even that they might disclose it to other entities. The debate is far-reaching and some people argue that there is no need for a CA at all.

In 1991, the first version of a software package called *Pretty Good Privacy* (*PGP*) was made freely available to anyone wishing to use strong encryption. It employed RSA for user authentication and symmetric key distribution and a symmetric encryption algorithm called IDEA for confidentiality. Although it used digital certificates, the original version of PGP did not rely on a central CA. Instead any user could act as a CA for anyone else. This became known as the *Web of Trust* approach. It essentially meant that users judge the trustworthiness of any certificate according to whether or not it is signed by someone they trust. For a small communications network, this type of approach certainly removes the need for a central CA and may work. However, there are a number of potential problems for large networks.

Another possibility for removing the need for a CA is to let a user's public key value be completely determined by their identity. If a user's identity and public key were (essentially) identical, then there would clearly be no need to have certificates to bind them together. The concept of identity-based public key cryptography was proposed by Shamir in 1984, and there have been a number of signature schemes based on this concept. However, it was not until 2001 that an identity-based public key algorithm for encryption was produced. There are two such algorithms; one due to Boneh and Franklin and the other designed at CESG (the UK's Communications and Electronic Security Group).

In identity-based systems, there has to be a universally trusted central body that computes each user's private key for their public key and delivers it to them. This approach does not therefore remove the need for a TTP, which generates every user's private key. Nevertheless it does remove the need for certificates. In this instance there is probably no advantage to user A claiming to be B, as only B has the private key determined by B's identity.

The use of identity-based public key systems represents an interesting alternative to the traditional PKI approach. Unfortunately it also presents its own problems. The most obvious relate to the concept of a unique identity and to the revocation of public keys. Suppose that a user's name and address were to determine their public key. If their private key were compromised, then they would have to move home or change their name. Clearly this is not practical. There are solutions to this particular problem of *identity theft*. One might be to let a user's public key depend on their identity and another publicly known variable, such as the date. This would ensure that a user's private key changed every day but might produce an unacceptable workload for the centre. There is considerable current research being conducted to see whether there are scenarios in which identity-based systems could be used to replace PKI.

At the other extreme, there are people who argue that the best way to approach security is to concentrate as many of the risks at a single location and then provide maximum security at that point. If this approach is adopted, then the CA may generate the users' keys. It is then often argued that if you trust the CA enough to generate your keys you might as well trust it to manage them on your behalf. The justification is that keys need the highly secure environment of the CA. This is known as the *server-centric* approach and is proving attractive to certain organizations.

Chapter 8
Key management

Introduction

In the earlier chapters, we concentrated on algorithms and their uses. However, we have repeatedly stressed the importance of good key management. In general, the effectiveness of cryptographic services depends on a number of factors that include the strength of the algorithm, a number of physical properties, including tamper-resistance of crucial hardware devices and access control to hardware, plus key management. Strong algorithms are used to prevent attackers calculating keys. However, their value is minimized if the attackers can obtain the appropriate keys by other means. The security of any cryptographic system is totally dependent on the security of the keys. The keys need protection during all phases of their life cycle. In this chapter we explain precisely what we mean by key management, look at the risks to which keys are likely to be exposed, and discuss some practical solutions. We often refer in this chapter to some of the more commonly adopted standards, notably those produced for the banking industry by the American National Standards Institute (ANSI). To be effective, a key management scheme needs to be carefully chosen to ensure that it meets the business needs and implementation requirements of the system. It must never be forgotten that overly elaborate cryptographic security systems represent a business overhead.

The key life cycle

The basic objective of key management is to maintain the secrecy and integrity of all keys at all times. For any given key this starts with the generation process and does not end until the key is no longer in use and has been destroyed. The main stages of a key's life cycle are shown in the diagram.

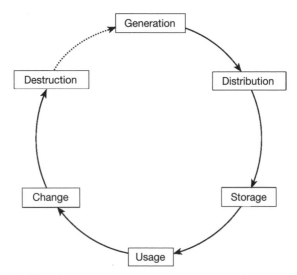

The key life cycle

In almost all situations each key is eventually replaced by another. Thus the process is a cycle in the sense that key destruction is followed by its replacement by a new key. However, this new key is likely to have been generated, distributed, and stored before the old key is destroyed. In some systems, there may be an extra requirement for key archiving.

Monitoring procedures are needed throughout any key's life, to detect potential attacks on the key. This almost certainly involves some form of audit trail to log usages of the key, but there is obviously no point having this audit trail unless it is monitored. Furthermore, the value of monitoring is vastly reduced unless someone has the authority to act in response to a potential key compromise. Thus it is often desirable, especially in large systems, for keys to have designated owners who are responsible for their protection.

We now discuss each element of the life cycle. Although many of the basic management principles are the same, the management of keys for symmetric cryptography is very different from that required for keys for asymmetric cryptography. In fact the establishment of a PKI is essentially the basis of certain aspects of key management for asymmetric algorithms. In our discussion, we concentrate on symmetric systems and include occasional comments to indicate when the difference between the two systems is significant.

Key generation

Key generation is often a problem, especially for public key algorithms where the keys have complicated mathematical properties. For most symmetric algorithms, any string of bits (or occasionally other characters) is likely to be a key. This implies that most users of symmetric algorithms have the capability to generate their own keys. The main problem is to generate the keys in such a way that they are unpredictable. Popular methods include manual techniques (e.g. coin-tossing), derivation from personal data (e.g. PIN) or a (pseudo-)random number generator.

For asymmetric systems, the situation is different. The generation of large prime numbers requires some sophisticated mathematical processing and may require considerable resources. As discussed in the previous section, users may be forced to trust externally generated keys or externally written software. If we consider RSA, its security relies on an attacker being able to factor the modulus N. If the key generation process gives only a limited number of primes, then the attacker could generate that limited number of primes and then try each prime as a factor. This is just one simple illustration of the importance of having good key generation for public key systems.

Key distribution and key storage

Both key storage and the distribution of keys are very important. The problems encountered and the solutions implemented to solve them tend to be similar, and so we discuss them together.

The reason for using a strong algorithm is to prevent attackers from being able to calculate the key. This is a pointless exercise if attackers can find it in clear somewhere in the system. Secure storage of certain keys is almost certain to involve some form of physical protection. For example, keys might be stored only in locations to which physical access is tightly controlled. The protection of those keys would then be solely dependent on the effectiveness of the access control. Alternatively, keys might be stored on a device such as a smart card for which there are two levels of protection. First, the owners are responsible for ensuring that they maintain possession of the card. Secondly, the card might have a high level of tamper-resistance to prevent people who obtain it from reading its contents.

A crude rule for key protection is that, ideally, keys should not appear in clear anywhere in the system unless they have adequate physical protection. If that physical protection is not available, then the keys should either be encrypted using other keys or split into two or more components. This rule was introduced at a time when most encryption was performed in hardware. If feasible, it is still sound practice as tamper-resistant hardware storage is regarded as providing more protection than software. The concept of having keys protected by encryption using other keys leads to the concept of a *key hierarchy*, where each key is used to protect the one beneath it. Key hierarchies are important and we discuss them later in this chapter. However, for the moment we merely note that it is clearly not possible to arrange a system so that every key is protected by another and there must be a key at the top of the tree. This *master key* is likely to be generated and distributed in component form. The components are separately owned and are installed separately into the encryption device. Clearly, for the concept of using components to be meaningful, it should not be possible for anyone to gain access to all components in clear form.

We now consider how key components can be constructed that do not give any information about the key. Suppose that we wish to have two components that can be combined to construct a key K. The naïve method would be to use the first half of K as the first component K_1 and the second half as the other component K_2. However, knowledge of component K_1 would enable the key K to be found merely by trying all possible values for the other component K_2. So, for instance, if K were a 64-bit key then knowledge of K_1 would mean that K could be determined in only 2^{32} trials, needed to find K_2, which is negligible when compared to the 2^{64} trials needed for an exhaustive key search for K. A far better solution would be to generate two components K_1 and K_2 of the same size as K such that the key K is the XOR of K_1 and K_2 ($K = K_1 \oplus K_2$). Since K and K_2 are the same size, knowledge of the component K_1 does not lead to a faster method for finding the key K, as it is no easier to search for K_2 than it is to search for K.

A more sophisticated approach is to use the concept of a *secret sharing scheme*. In this type of scenario there are a number of values, known as shares, and the key is obtained by combining some, or all, of them. One possibility might be, for instance, to have seven shares and design the system so that any four of those shares determine the key uniquely but knowledge of any three gives no information about the key. This not only introduces the security associated with shared responsibility but places less reliance on the availability of specific individuals if key recovery becomes necessary.

As with many other aspects of cryptography, key management for communication systems is considerably harder than for the handling of stored data. If a user is merely protecting their own information, then there may be no need for any distribution of keys. However, if there is a need to communicate securely, then there is often a need for key distribution. Furthermore, the magnitude of the associated problem is likely to depend upon the number of terminals that are trying to communicate securely. If there are only two such terminals, then this is referred to as a *point-to-point* environment. If there are more than two communicating terminals, then the solution to the key distribution problem is likely to depend both on the business application and the environment formed by the terminals. There are two extremes. One is the 'hub and spoke' environment, which consists of a central terminal (or hub) and a number of other terminals that must be able to communicate securely with the hub. The other is a 'many-to-many' environment that occurs where each terminal may require a secure link to every other.

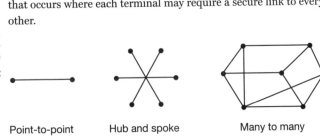

Point-to-point   Hub and spoke   Many to many

For public key systems the situation is different. Much of this discussion applies to private keys as, like symmetric keys, they need to be kept secret. However, public keys are stored and distributed in certificates, as discussed in Chapter 7.

Key establishment

The idea of key establishment is that two parties have a method for agreeing upon a key between themselves. Such a method is called a *key establishment protocol* and is an alternative to distributing keys. It is, of course, crucial that the two parties are able to authenticate each other prior to establishing this key. The use of public key certificates makes this possible. The most well-known and widely used protocol of this type is due to Diffie and Hellman. In the *Diffie–Hellman protocol*, the two parties exchange their public keys. Using a carefully chosen combining rule, each party then combines their own private key with the other's public key. This gives them a common value from which the key is derived.

The need for the two users to be able to authenticate themselves is crucial. Without it, the protocol is subject to the so-called 'man-in-the-middle attack'. In this attack, an attacker intercepts communications between the two parties and then impersonates each party to the other. The result is that the two parties believe they have an agreed key with each other but, in fact, each party has agreed a key with the 'man in the middle'. This is a situation where digital certificates are invaluable.

The basic idea of the Diffie–Hellman key establishment protocol is that, even if they manage to eavesdrop on the key establishment communications, interceptors are not able to calculate the key. Quantum cryptography is an interesting new key establishment technique which does not rely on the strength of a cryptographic algorithm. The properties of quantum mechanics are used by the two parties for both the transmission of information and detection of interception of that transmission. The establishment of a key involves one user sending a random sequence to the other. If it is intercepted then that interception can be detected, and the establishment process is recommenced. A non-intercepted sequence is then used as the basis for the key.

Key usage

In many systems, each key has a prescribed usage and must be used for that purpose only. It is not clear that this requirement is always justified. However, there have undoubtedly been instances where weaknesses have resulted from multiple uses of single keys. It is now considered good practice to maintain separation of use.

We have already seen examples that demonstrate that the concept

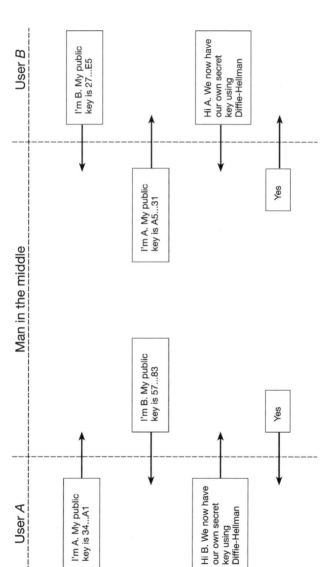

Man in the middle

of using keys for only one purpose is a good idea. For example, we have already discussed the use of keys for encrypting other keys, which is regarded as distinct from encrypting data. In order to understand such a restriction of use in practice, we need to reintroduce the concept of a *tamper-resistant security module* (TRSM). If a user receives some ciphertext, then with this ciphertext and the appropriate key as input to the TRSM, the user expects the TRSM to output the data. However, if the user receives an encrypted key, then the user does not want the clear key as output from the module. Instead the user wants the key to be decrypted and used within the module. However, the ciphertext and the key are both bit-strings and the encryption algorithm cannot distinguish between them. Thus, the concept of key usage must relate to the functionality of the TRSM and not to the actual algorithm.

In order for keys to have unique usages, we need to assign a label to each individual key. This label specifies the key's use. For example, this label might specify a 'data encrypting key', 'key encrypting key', 'MAC generation key', or 'MAC verification'. Of course, the exact form of these labels depends on the TRSM and on the environment. For an asymmetric algorithm, users may need two public and private key pairs, one pair to be used for encryption and the other for signatures.

Once labels have been agreed, there must be a method of binding the label to the key so that adversaries cannot change the label and thereby misuse the key. One method is to let all encrypted versions of the key depend on the top-level key of the TRSM and the key's label. This ensures that the label could only be 'removed' within the TRSM. Once labels have been bound to keys, there has to be some mechanism for ensuring that keys cannot be misused. The design and configuration of the TRSM are crucial for enforcing this mechanism.

Key change

In all cryptographic systems, there must be the ability to change keys. There are many reasons for this, and changes may be either scheduled regular updates or responses to suspected compromise. If it is suspected that a key has been compromised, then it should probably be changed immediately. Many organizations have regular key changes as dummy runs so that they are prepared for an emergency and their staff have the relevant practical experience.

Keys are changed regularly to limit their exposure and reduce their value to an attacker. The value of a successful attack clearly determines the time and effort an attacker is likely to invest. There are EFTPOS (Electronic Funds Transfer at Point of Sale) systems where the key is changed after every single transaction. For such systems, an attacker is unlikely to invest large resources to launch an attack which yields only a single key and one compromised transaction.

There are no clear rules about how frequently keys should be changed. However, it is clear that every key should be changed long before it could be determined by an exhaustive key search. Another factor might be the perceived risks of the key being compromised balanced, on the other hand, by the risks associated with changing the key.

Key destruction

Keys must be destroyed in a secure manner whenever they are no longer needed. Thus, simply deleting a file that stores the key value is not sufficient. It is often desirable to provide meticulous detail of how it should be done. For instance, a relevant ANSI Standard states: 'Paper-based keying materials shall be destroyed by crosscut, shredding, burning or pulping. Key material stored on other media shall be destroyed so that it is impossible to recover by physical or electronic means.' In particular, this means that all electronically-stored keys should be positively overwritten on deletion, so that they do not leave a trace or other information that could be useful to an attacker. This is particularly important in software applications where the memory that is used to store keys may be used for some other purpose later on.

Key hierarchies

As we have already noted, manual processes are both time-consuming and expensive. It is clearly desirable to restrict them to a minimum. Thus for key management, where possible, electronic key distribution is preferred to manual distribution. However, if a key is distributed electronically, then it needs protection from exposure during transmission. The favoured way of achieving this is to encrypt the key with another key. As has already been noted, this leads to the concept of a key hierarchy where the top-level or master key has no key to protect it. Thus the master key needs to be distributed manually, either in a tamper-resistant device or in component form.

The simplest form of key hierarchy is one with two layers. The master key is a key-encrypting key, and is used solely to protect keys on the lower level. The lower level keys are called either *session keys* or *working keys*. Their function varies according to the application. For example, they may be used to encrypt data for confidentiality or to calculate MACs for data integrity. The session may be defined in a number of ways, possibly in terms of a time duration or a number of uses. When the session key needs changing, the new replacement is distributed under the protection of the master key. However, if the master key is to be changed then that process needs to be handled manually. Manual key changes are often impractical and so many systems have three-layer hierarchies, with an extra level between the master key and the session key. Keys in this layer are key-encrypting keys and their role is to protect the session keys. However, these key-encrypting keys can now be distributed under the protection of the master key. This extra layer enables the change of key-encrypting keys electronically and considerably reduces the likelihood of having to change keys manually. The two options are

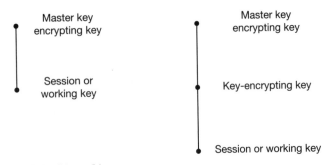

described in the above diagram, where each key 'protects' the one beneath it.

Throughout this discussion of key management, we have been assuming that the working keys are symmetric. However, there is no reason why the algorithm used to encrypt working keys should be the same as the algorithm used to protect data. In particular, the fact that the working keys are symmetric does not prevent the use of public key cryptography for the top-level keys. In fact there are many *hybrid* systems where asymmetric algorithms are used to distribute keys for symmetric algorithms.

Managing keys in networks

If two parties wish to exchange encrypted messages, then there are a number of options open to them for handling keys, depending on their environment and the level of security that is required. The users could meet to exchange key values physically. If the users agree to use an encryption product, then that product may facilitate the key agreement for them by using a protocol such as the Diffie–Hellman protocol. However, a product may be too expensive or complex. For example, there are products which provide an encryption algorithm and give advice to the user on constructing a key as a string of alphanumeric characters. The sender is then advised to pass the key to the receiver by means of a telephone call. The security implications are clear, but are probably acceptable for most personal communications. Nevertheless, most people do not want the hassle of having to make a phone call in order to send a confidential email message.

If the level of security required is high, then the initial key agreement is likely to involve some form of manual process. Since such processes tend to be slow and expensive, users are likely to try to ensure that all future key agreements can be achieved electronically. If a network is small enough, then one option for the key distribution is for each pair of terminals to establish a shared key between themselves. However, this can be time-consuming and expensive. For a large network, there is a danger of key management becoming too great an overhead. In order to overcome this problem, many networks have trusted centres whose roles include the facilitation of key establishment between pairs of users in the network.

One typical scenario involves each user establishing a shared key with a trusted centre. Although this may be time-consuming and expensive it has to be done only once. If two users then wish to communicate privately, they request the assistance of the trusted centre to establish a shared secret key, using the secret keys which already exist between the trusted centre and each of the users. The solutions which we discuss are based on ANSI and ISO standards.

Using a trusted management centre

The scenario we consider is that of a large network in which each node may require a secure cryptographic link with any other node. The size of the network necessitates the use of a trusted centre to facilitate the establishment of a shared secret key between any two nodes. We assume that each node has established a permanent secure link with the centre. Any two nodes may then seek the assistance of the trusted centre to establish a shared secret key.

Although the two nodes use a symmetric algorithm for their secure communications, the algorithm used for secure communications between the centre and each node may be either symmetric or asymmetric. If a symmetric algorithm is used throughout the system, then the trusted centre is either a *Key Distribution Centre* or a *Key Translation Centre*. If an asymmetric algorithm is used between the trusted centre and the nodes, then the centre is a *Key Certification Centre*. We discuss each situation in turn.

Suppose a symmetric algorithm is used throughout the system. If node A wants to have a secure communication with node B, then A may request the trusted centre's help in establishing a shared secret key between A and B. With a Key Distribution Centre, A asks the Key Distribution Centre to provide the key; whereas, with a Key Translation Centre, A generates the key and asks the Key Translation Centre to enable it to be distributed securely to B. In either case, the keys which A and B share with the centre can be used as key-encrypting keys to protect all communications between either node and the centre. If we denote the new key by KAB, then whenever KAB is transmitted it is protected by one of the keys which the nodes share with the centre. Thus both nodes A and B are relying on the secrecy of the keys they share with the centre for their confidence in believing that the only nodes that know KAB are A and B.

Suppose now that an asymmetric scheme is used between the trusted centre and the nodes. We assume that nodes A and B wish to communicate and have public and private key pairs. We further assume that the Key Certification Centre knows these public key values and is able to assure each of nodes A and B of the authenticity of the other's key. The simplest method is probably for the KCC to act as a Certification Authority and issue certificates binding nodes A and B with their respective public keys. Suppose node A now generates a symmetric key KAB for secure communication with B. Node A can encrypt symmetric key KAB with node B's public key and then sign the result with node A's own private key. Encrypting

KAB with node B's public key gives node A confidence that the symmetric key KAB is also known only to node B. Furthermore, signing KAB with node A's private key gives node B confidence that the symmetric key KAB must have originated at node A. Thus, both nodes A and B are sure that they are the only two nodes that know the symmetric key KAB.

The symmetric key shared between A and B may now be used either as a key-encrypting key or a working key. If KAB is a key-encrypting key, then nodes A and B need never use the trusted centre again to establish working keys. Furthermore, a Key Certification Centre cannot calculate the symmetric key KAB if A and B generate their own public and private key pairs. However, with either a Key Distribution Centre or Key Translation Centre, the symmetric key KAB must appear in clear at the trusted centre.

Key recovery and key back-up

Anyone wishing to obtain the plaintext corresponding to some ciphertext needs at least one of the following to occur:

1. They are given the plaintext.
2. They know the decryption algorithm and are given the decryption key.
3. They know the decryption algorithm and have the ability to break it.
4. They can locate the plaintext somewhere within the system.
5. They know the decryption algorithm and can locate the decryption key within the system.
6. They have the ability to deduce the algorithm and break it.

If situation 1 occurs then they can bypass the cryptography while situation 2 gives them the same information as the intended receiver. The use of strong encryption is to thwart attacks appropriate to situation 3. However, the use of strong encryption is pointless in either situation 4 or 5. If situation 4 applies then they can bypass the cryptography, while situation 5 means that they have the same knowledge as the genuine receiver without breaking the algorithm. It is, therefore, very important that keys are secured during their complete life cycle. We have already discussed key management in detail, but have not yet mentioned the important issue of key back-up. It is important to realize that vital information may be lost forever if it is encrypted using a strong algorithm and the key is subsequently lost or corrupted. Thus it is important to have back-up copies of the key, which are securely stored locally or at a trusted third party. We are assuming the worst case conditions described earlier, so we do not consider situation 6.

When discussing encryption we have usually adopted the stance that it is a tool that is used either by individuals or companies to protect their private communications or stored information. It also, of course, provides criminals and terrorists with similar protection from law enforcement and other government agencies. For many years law enforcement agencies have argued that the interception of communications is vital in the fight against crime. In recognition of this, many countries have long-standing legislation which, under certain conditions, permits the legal interception of certain communications, such as telephone calls. Similar assertions are made by intelligence services in order to counter terrorism and other threats to national security. Different countries have reacted to these problems in different ways. Some governments tried to maintain strict control over all uses of encryption, while others, including the USA and the UK, restricted their control to the export of encryption devices. However recent developments, most notably the rapid expansion of the use of software encryption algorithms, have caused most governments to rethink their policies on the use of encryption.

There is an obvious conflict of interest between private individuals and organizations, who want to protect their confidential data, and law enforcements agencies, who assert that they need to be able to read specific intercepted traffic to fight crime and to protect national security. Companies want encryption that is strong enough for organized crime not to be able to break it, while governments want to be able, in certain circumstances, to access any transmission.

The UK's Regulation of Investigatory Powers Act 2000 is concerned with the interception of communications. Not surprisingly, the section relating to lawful interception has caused considerable debate and controversy. Part of that controversy surrounds the stipulation that, under certain restrictions, law enforcement agencies may demand either the cryptographic key needed to decrypt an intercepted cryptogram or be given the corresponding cleartext.

Certainly much of the debate concerns the moral issue of whether law enforcement agencies have the right to demand keys under any circumstances. This is a particularly modern example of the debate about the balance between the rights of the individual and the needs of the state. We intend to take no position in this book on that particular issue. However, we do observe from a technical perspective that a user who accepts that law enforcement agencies are entitled to read encrypted data under certain conditions may find it to their advantage if only situations 1 or 2 of the six listed above can occur. If any of the situations 3 to 6 can occur for law enforcement agencies, then they are also likely to occur for a sufficiently well-resourced adversary.

The use of encryption by individuals to provide confidentiality for communications such as emails is not as widespread as had been anticipated. This is certainly not because of lack of available algorithms. Indeed would-be users are almost spoilt for choice of publicly available algorithms that have been subjected to open academic scrutiny and appear to be very strong. The main reason is probably the lack of easily used products. Most individuals are not concerned enough about security that they are prepared to subject themselves to any extra effort to obtain it. When sending an email, a user normally just wants to be able to click the 'send' button. However, a request for the use of encryption usually activates a sequence of questions from the computer which expect answers or actions from the user. Most users cannot be bothered. Much of the hassle associated with using encryption comes from handling the keys. Unfortunately, as we have repeatedly stressed, good key management is also absolutely crucial for the overall security of the system.

Chapter 9
Cryptography in everyday life

Introduction

Throughout the text we have repeatedly stressed the relevance of cryptography to modern life and have used real-life situations to illustrate some of the important issues. This chapter contains a number of disjointed situations where the use of cryptography facilitates the provision of a secure service. Many of them represent scenarios that the man in the street encounters almost every day, but probably does not appreciate either the security risks or the role played by encryption. In each case we describe the application, discuss the security issues, and show how cryptography is used.

A cash withdrawal from an ATM

When someone makes a cash withdrawal from an Automated Telling Machine (ATM), they need to produce a plastic, magnetic stripe card and have knowledge of the associated PIN. The customer places their card in the ATM slot and enters their PIN. They then enter the amount requested for withdrawal. In a typical transaction the system needs to check that the PIN is the correct one for that card and, if the transaction is online, that the customer is allowed to withdraw the requested amount. This verification is likely to take place at the bank's central computer and therefore there must be two-way communication between the ATM and the host computer. The ATM sends the card details and PIN to the host computer, and the response message either authorizes the transaction or refuses it. Clearly these communications need protection.

Although the amount of a withdrawal may not be secret, it is important that the amount dispensed at the machine is identical to the amount debited from the account. Thus the message needs some form of integrity protection. Furthermore banks are understandably nervous about the possibility of an ATM paying out on the same positive response message more than once. Thus there is also a requirement to include sequence numbers on the response messages to prevent replays.

All banks instruct their customers to keep their PINs secret as anyone who knows the correct PIN can use a stolen or lost card. Clearly the banks must ensure that the PIN is not compromised within their system and so the PIN is encrypted during transmission and on the database that is used for checking the validity of the PIN. The algorithm used for this process is DES in ECB mode. Since DES encrypts 64-bit blocks and PINs are, typically, only four digits, the block containing the PIN needs padding before it is encrypted. If the padding were identical for all customers then, even though they do not have the correct key, anyone who gains access to encrypted PIN blocks would be able to identify customers who share the same PINs. Using a padding technique where the padding depends on customer card details eliminates this potential weakness.

This use of encryption prevents the PIN being exposed to eavesdroppers who intercept the communications between the ATM and the host computer. They also prevent PINs from being read by personnel who have access to the bank's database. However, as we discussed earlier, encryption cannot prevent a fraudster guessing someone's PIN. Anyone who finds or steals a plastic card can enter it into an ATM and try a lucky guess. Since there can be at most 10,000 four-digit PINs, the chances of a successful guess are not ridiculously small. In recognition of this, most ATMs allow only three PIN tries before they 'swallow' the card. This is seen as a reasonable compromise between the security risk of allowing fraudsters too many guesses and the risk of inconveniencing genuine cardholders who make mistakes when entering their PINs. As has already been noted, the use of encryption cannot protect against guessing the PIN.

Some ATM networks now use smart cards, which enable the use of public key cryptography. A user's card then contains their private key and a certificate, signed by the card issuer, to confirm their public key value. The ATM authenticates the card by issuing a challenge for the card to sign. As with all systems that rely on certificates, it is necessary for the terminal to have an authentic copy of the card issuer's public key in order to check the validity of the certificate. For some systems this is achieved by having its value installed into the ATMs.

Pay TV

Anyone who subscribes to a Pay TV system expects to able to view those programmes for which they pay and may also expect that people who have not paid do not have access to those programmes. Pay TV systems are one example of a controlled access broadcast network. In such a network information, in this case the TV programme, is broadcast widely but only a restricted set of those who receive the signal can actually understand the information. A common way of achieving this objective is for the broadcast signal to be encrypted using a key that is made available only to the intended recipients of the information. There are many ways of establishing and managing such systems.

In a typical Pay TV system each programme is encrypted with its own unique key prior to transmission. Those who pay for a particular programme are then essentially paying for knowledge of the key. Clearly this creates a key management problem, namely that of being able to deliver the key to the correct viewers. A common solution to this problem is to issue each subscriber to the network with a smart card that contains that subscriber's unique private key for an asymmetric encryption algorithm. This smart card is then inserted into a reader which is either part of the TV or in an attachment provided by the TV network operator. When a subscriber pays for a particular programme, the symmetric key used to encrypt that programme is transmitted encrypted with the subscriber's public key. Thus, using the language of Chapter 8, this type of system is using a two-level key hierarchy with a hybrid of symmetric and asymmetric algorithms.

Pretty Good Privacy (PGP)

Pretty Good Privacy or PGP was originally developed by Phil Zimmermann in the late 1980s. It was intended as a user-friendly product for encryption on a personal computer and uses both symmetric and asymmetric cryptography. Many versions of it are now being used. We discuss the general concept without concentrating on any particular version or applications.

PGP uses a two-level key hierarchy in which symmetric session keys are used to protect data, and asymmetric keys are used for both signature and the protection of the symmetric session keys. PGP has many uses including securing email and the secure storage of files. The publication of PGP on a bulletin board in 1991 brought Phil Zimmermann into dispute with both the US government (for the alleged illegal export of cryptography) and various patent holders. These disputes had been resolved by 1997. PGP is available as freeware and is included with the software of many new PCs.

As we have already discussed, a major problem in using asymmetric encryption is that of key authentication. We have discussed one solution to this problem of using a network of Certification Authorities (CAs) in a Public Key Infrastructure (PKI). PGP introduced a different solution to the problem of public key authentication: the *web of trust*.

A web of trust might be established as follows. Initially, each user signs their own public key as being authentic, that is, the user essentially acts as their own Certification Authority. Suppose now that users A and B each have such self-signed keys. If user B 'trusts' user A, then B would be happy to sign A's public key as authentic. So user B is essentially acting as a CA for user A. Suppose now a user C does not know user A, but wishes to be certain that A's public key is authentic. If C 'trusts' any user who has signed A's public key, then C is happy that user A's public key is authentic. This user is known as an *introducer* of A to C. By such cross-signing of public keys, a large intricate network of authenticated public keys (the web of trust) can be built up. This allows a user to associate a level of trust with each public key depending on that user's perception of the trustworthiness of signatories to that public key's authenticity.

There have been many versions of PGP since its introduction in 1991, the latest (2001) being version 7. Earlier versions of PGP used RSA and IDEA for the asymmetric and symmetric cryptographic algorithms; whereas later versions of PGP used (by default) Diffie–Hellman/El Gamal and CAST for their asymmetric and symmetric cryptographic algorithms. We now briefly outline the cryptographic processes performed by the various PGP options as used for email.

PGP keys

This option displays a window listing all stored asymmetric key pairs of the user and all stored public keys of other users, together with a level of trust and a list of signatures associated with each of these keys. There are facilities in this window to authenticate and sign other users' public keys, and to export and import public keys with their signatories. This option also permits a user to generate new asymmetric key pairs based on data derived from mouse movements and keystrokes. The private key of a user's key pair is then stored encrypted using a symmetric cryptographic algorithm and a user-chosen *passphrase* or key.

Encrypt

This option encrypts the message using a symmetric encryption algorithm with a session key based on data derived from mouse movements and keystrokes. This session key is encrypted using the public key of the recipient. The encrypted message and the encrypted session key can then be sent to the recipient. The recipient can use their private key to recover the symmetric session key and hence the message.

Sign

This option signs the message using the sender's private key. The recipient can check this signature using the sender's public key.

Encrypt and Sign

This option both signs and then encrypts the message as outlined above.

Decrypt/Verify

This option allows the receiver to decrypt an encrypted message or to verify a signature (or both).

Secure web browsing

Many people now shop on the web. To do so they almost certainly use a credit card, which means that their credit card details are transmitted across the Internet. Concerns about the security of these details is often listed as one of the main reasons why this form of shopping is not more widely used. In this short section we discuss how credit card details are protected on the web and extend the discussion to other security issues.

Secure web browsing is an essential feature of e-commerce. The *Secure Sockets Layer* (*SSL*) and the *Transport Layer Security* (*TLS*) are two important protocols that are used to verify the authenticity of websites. They facilitate the use of encryption for sensitive data and help to ensure the integrity of information exchanged between web browsers and websites. We concentrate on SSL.

SSL is an example of a client-server protocol where the web browser is the client and the website is the server. It is the client that initiates a secure communication, while the server responds to the client's request. The most basic function that SSL is used for is the establishment of a channel for sending encrypted data, such as credit card details, from the browser to a chosen site.

Before discussing the protocols we note that web browsers typically hold some cryptographic algorithms together with the public key values of a number of recognized Certification Authorities.

In the initial message from browser to site, often referred to as the 'client hello', the browser has to send the server a list of the cryptographic parameters that it can support. However, although it initializes an exchange of information that enables the use of encryption, that message does not identify the browser to the site. In fact, for many applications, the site is not able to authenticate the browser and the authentication protocol merely authenticates the site to the browser. This often makes sense. If, for instance, an individual wishes to make a purchase using the web browser then it is very important for them to establish that the website they are browsing is authentic. The merchant, on the other hand, may have other means of authenticating the user's identity or may not even care about it. Once the merchant has received a credit card number, for example, it can validate that number directly with the card issues.

The website authenticates itself to the browser by sending its public key certificate which, provided the browser has the appropriate public key, provides the browser with an authentic copy of the site's public key. As part of the establishment of the secure channel, the browser then sends the site a session key for an agreed symmetric algorithm. The session key is encrypted using the site's public key, thereby giving the browser confidence that only the nominated site can use it. Thus SSL provides another everyday example of the type of hybrid key management system discussed in Chapter 8. It also provides an example of the use of a PKI for entity authentication.

Using a GSM mobile phone

One of the main attractions for users to have mobile phones is that they offer the ability to roam and to make telephone calls from almost anywhere. However, since the mobile phones are wireless, the phone message is transmitted across the airwaves until it reaches the nearest base station, where it is transferred to the fixed landline. Since intercepting radio signals is likely to be easier than intercepting landline calls, one of the initial security requirements for GSM was that their mobile phones should be no less secure than the conventional fixed telephones. This requirement was satisfied by providing encryption for transmissions from the handset to the nearest base station. Another serious security issue was the problem of the operator being able to identify the phone so that they knew whom to charge. Thus, for GSM, there were the following two major security requirements: confidentiality, which was a customer requirement and user authentication, which was a system requirement.

Each user is issued with a personalized smart card, called a SIM, which contains a 128-bit secret authentication value known only to the operator. This value is then used as the key to a challenge–response authentication protocol using an algorithm, which can be selected by the operator. When the user attempts to make a call, their identity is transmitted to the system operator via a base station. Since the base station does not know the SIM's secret key, and may not even know the authentication algorithm, the central system generates a challenge and sends it, with the response appropriate for the card, to the base station. This enables the base station to check the validity of the response.

In addition to the authentication algorithm the SIM also contains a stream cipher encryption algorithm, which is common throughout the network. This algorithm is used to encrypt the messages from the mobile phone to the base station. The key management for the encryption keys is ingenious and makes use of the authentication protocol. The authentication algorithm accepts a 128-bit challenge and computes a 128-bit response, which depends on the card's authentication key. However only 32 bits are transmitted from the SIM to the base station as the response.

This means that when the authentication process has been completed there are 96 bits of secret information known only to the SIM, the base station, and the host computer. Of these bits, 64 are then allocated for the determination of the encryption key. Note that the encryption key changes every time that authentication takes place.

References and further reading

We now give complete references for the books referred to in the text and provide a few suggestions for further reading. As we have already noted, there are a very large number of books written on most aspects of cryptography and, consequently, we have made no attempt to provide a comprehensive list.

Throughout the text we have referred to Menezes, van Oorschot, and Vanstone, *Handbook of Applied Cryptography*, as our standard reference for more detailed discussions of technical cryptographic issues. This book is highly regarded by the academic community and by practising cryptographers. We recommend it for anyone who intends to study cryptography seriously. However, it is important to point out that, as with most other cryptographic textbooks, it assumes that the reader has a sound mathematical background. For anyone who does not have sufficient mathematical training, but would like more technical details than we have provided, we suggest R. E. Smith, *Internet Cryptography*. (Cooke, *Codes and Ciphers*, is included for parents who want to interest their (very) young children!) If you want more exercises then try Simon Singh's website. It includes interactive encryption tools, cryptograms to crack, and a virtual Enigma machine.

For security professionals, digital signatures and PKIs are likely to be the most relevant aspects of cryptography. These two topics are covered in the works listed by Piper, Blake-Wilson and Mitchell, and by Adams and Lloyd, respectively. Readers who want to see how cryptography 'fits' into the wider concept of securing electronic commerce should consult Ford and Baum, *Secure Electronic Commerce*. For those who wish to see cryptography in a wider setting, we recommend Anderson, *Security Engineering*.

As we noted in Chapter 1, the history of cryptography is a fascinating subject. The 'classic' tome on the subject is Kahn's *The Codebreakers*, while Singh's *The Code Book* is a more recent book that has been highly influential in increasing public awareness and enjoyment of cryptography. One particular historical event that has been the topic of many books, plays and films is the code-breaking activity at Bletchley Park during the Second World War. The first account is contained in Welchman, *The Hut Six Story*. *Codebreakers* (ed. Hinsley and Stripp) is a collection of a number of contributions from individuals who took part in the Bletchley Park saga. The events at Bletchley Park have been made into a highly successful film based on the novel by Robert Harris.

Throughout history there has been conflict between individuals/ organizations who wish to protect their confidential information and governments who have tried to control the use of encryption. This conflict is discussed in Diffie and Landau, *Privacy on the Line*.

The other references are included to provide source material for various facts or topics included in the text.

Carlisle Adams and Steve Lloyd, *Understanding Public-Key Infrastructure* (Macmillan Technical Publishing, 1999)

Ross Anderson, *Security Engineering* (John Wiley & Sons, 2001)

Henry Beker and Fred Piper, *Cipher Systems* (Van Nostrand, 1982)

Guy Clapperton (ed.), *E-Commerce Handbook* (GEE Publishing, 2001)

Jean Cooke, *Codes and Ciphers* (Wayland, 1990)

W. Diffie and M. Hellman, 'New Directions in Cryptography', *Trans. IEEE Inform. Theory*, (Nov. 1976), 644–654

Whitfield Diffie and Susan Landau, *Privacy on the Line* (MIT Press, 1998)

Warwick Ford and Michael S. Baum, *Secure Electronic Commerce* (Prentice Hall, 1997)

Robert Harris, *Enigma* (Hutchinson, 1995)

F. H. Hinsley and Alan Stripp (eds.), *Codebreakers* (OUP, 1994)

B. Johnson, *The Secret War* (BBC, 1978)

David Kahn, *The Codebreakers* (Scribner, 1967)

Alfred J. Menezes, Paul C. van Oorschot, and Scott A. Vanstone, *Handbook of Applied Cryptography* (CRC Press, 1996)

Georges Perec, *A Void*, tr. Gilbert Adair (Harvill, 1994)

Fred Piper, Simon Blake-Wilson, and John Mitchell, *Digital Signatures: Information Systems Audit and Control* (Information Systems Audit & Control Association (ISACA), 2000)

C. E. Shannon, 'A Mathematical Theory of Communication', *Bell System Technical Journal*, 27 (1948), 379–423, 623–56

C. E. Shannon, 'Communication Theory of Secrecy Systems', *Bell System Technical Journal*, 28 (1949), 656–715

Simon Singh, *The Code Book* (Fourth Estate, 1999)

Richard E. Smith, *Internet Cryptography* (Addison Wesley, 1997)

Vatsyayana, *The Kama Sutra*, tr. Sir R. Burton and F. F. Arbuthnot (Granada Publishing, 1947)

Gordon Welchman, *The Hut Six Story* (McGraw-Hill, 1982)

Websites

http://www.cacr.math.uwaterloo.ca/hac/ *Handbook of Applied Cryptography* website

http://www.simonsingh.com/codebook.htm *The Code Book* website

http://www.rsasecurity.com/rsalabs/faq/ RSA Laboratories' 'Frequently Asked Questions'

http://csrc.nist.gov/encryption/ National Institute of Standards (NIST) cryptography website

http://www.esat.kuleuven.ac.be/~rijmen/rijndael/ Rijndael (AES) website

http://www.iacr.org International Association for Cryptologic Research (IACR) website

Index

Cryptography